150 YEARS PHYSICS based on the WRONG EQUATION

150 YEARS OF PHYSICS BASED ON THE WRONG EQUATION

Light contains the key to open the doors to Heaven.
Unfortunately, the same key fits on the doors to Hell

Author: Wim Vegt

Country: The Netherlands
Website: https://wimvegt.topworld.center
Email: j.w.vegt@topacademy.center

1

Books from Wim Vegt in the series: "The POWER OF LIGHT":

10) The Nikola Tesla Way of Energy Transport. (E-book) ISBN: 9789402191349. Paperback ISBN: 9789402190984.

9) The Rise of ELF Electromagnetic Attack Weapons and the Necessity of the Development of Corresponding ELF Defense Systems. (E-book) ISBN: 9789402189544. Paperback ISBN: 9789402189117

8) Unified 4-Dimensional Hyperspace Equilibrium. (E-book) ISBN: 9789402181036. Paperback ISBN: 9789402180985

7) Beyond Superstrings. (E-book) ISBN: 9789402179668. Paperback ISBN: 9789402179637

6) The Hidden World Behind Superstrings. (E-book) ISBN: 9789402180053

5) Light is the Bridge between God, Relativity and Quantum Physics (E-book)
ISBN: 9789402178975

4) The Particle-Wave-Mass Unification. A New Theory in Quantum Physics.
(E-book) ISBN: 9789402178647. Paperback ISBN: 9789402178586

3) The Tri-Unity in Religion and in Science. (Paperback) ISBN: 9789402178531

2) The Power of the LIGHT rules over the SHADOWS of the DARKNESS (Paperback) ISBN: 9789402178326

1) The Bridge of Light (E-book) ISBN: 9789402177947. Paperback ISBN: 9789402177763

Index

5

1.0 Introduction

When we look at todays Physics, we can only be impressed by an enormous amount of knowledge and a complete New World of technical applications that has never been in the world before. We now live in the century of the impressive victory of the new science and the new technology over the old-fashioned world and the old-fashioned way of thinking.

Great changings in the way of thinking and the technological achievements are mostly characterized by an important scientific publication in a century that changes everything in that century. We can recognize the century of Isaac Newton who triggered in 1687 the large changings in thinking with his famous publication "Philisophiae Naturalis Principia Mathematica" (Mathematical Principles of Natural Philosophy).

We recognize the century of James Clerk Maxwell who triggered in 1865 the large changings in thinking with his famous publication "A Dynamical Theory of the Electromagnetic Field".

We recognize the century of Albert Einstein who triggered in 1905 the large changings in thinking with his famous theory of Special Relativity represented in his publication "On the Electrodynamics of Moving Bodies". Manifesting a "New Theory" and a "New Way of Thinking" with important contributions of Hendrik Lorentz, Henri Poincaré and Hermann Minkowski.

It is recognizable that with the suddenly changing in thinking in a new period, a new kind of mutual common sense and a general agreement by many scientists of the the new theory and the new way of thinking rises. The new theory becomes like a medieval town with a large high wall around it. The New Theory will be protected by common sense and mutual agreement.

This new way of thinking settles down in the scientific society and become immovable. Other options disappear and simply do not exist anymore.

Different from the alpha and the gamma sciences, the beta sciences are being developed by a kind of a LEGO system. Building blocks built one after another and built on top of each other. Like we build with the LEGO system houses and castles using the same LEGO building block over and over again, we build in the beta sciences grand theories, using basic the same basic equations over and over again. A large shift in the beta sciences happens when a new mathematical building block has been developed. Like the equations of Newton or the equations of Maxwell or the Schrödinger and the Dirac equations. These fundamental equations form the mathematical LEGO system of our modern scientific world.

Because these mathematical building blocks are being used over and over and again in numerous applications over a period of of more than 100 years, a general scientific common sense rises around these mathematical building blocks. This scientific common sense protects these mathematical building blocks like a high wall around a medieval town.

A fundamental problem rises when one of these building blocks is not correct or turns out not to be correct under certain conditions. Like the famous Law of Newton for the relationship between acceleration (a), mass (m) and force (F): "$F = m\,a$" turns out not to be valid at velocities near the speed of light because at these velocities the mass is changing. It took a long time before Albert Einstein's theory of general relativity had been accepted, because his theory of general relativity was in contradiction with the famous well-known mathematical building blocks which had already been used and being protected for hundreds of years. But nowadays Einstein's famous theory of general relativity has been accepted world-wide.

7

This book describes a comparable conflict in the modern beta sciences and brings the well-known and generally accepted Modern Physics of the last 150 years in conflict. Because when a fundamental mathematical building block, which has been introduced 150 years ago and has been used to develop the Modern Physics during the last 150 years, turns out to be wrong (or not complete), a fundamental problems rises in Modern Physics, developed during the last 150 years.

Because when one of the many mathematical building blocks turns out to be wrong, the whole physics which has been built by using all these mathematical building blocks together might be wrong or not complete.

This situation happens in relation with the well-known Maxwell Equations, presented 150 years ago in the famous publication: "*A Dynamical Theory of the Electromagnetic Field" in 1865.* which has been used as a fundamental mathematical building block in many modern physical theories.

In Maxwell's time there were no optical LASERS (**L**ight **A**mplification by **S**timulated **E**mission of **R**adiation) and the outcome of his theory was in his time completely in correspondence with what could be measured at that time. The value for the speed of light, calculated from the Maxwell Equations, corresponded almost exactly with the value for the speed of light measured in 1862 by Léon Foucault by a system of rotating mirrors and measured in 1877 by Albert Michelson (300.140 [km/s]).

But nowadays there rises a problem with Maxwell's theory for the electromagnetic field. Since the existence of the LASERS it became clear that the speed of light is not always the same in every direction. When a beam of light, generated by a LASER, propagates with the well-known speed of light "$c = 299.792$ [km/s]" in the z-direction, the speed of light equals zero in the x-direction and the y-direction (in a orthogonal x,y,z frame).

This new phenomenon cannot be explained by Maxwell's Theory. In Maxwell's Theory the speed of light has to be exactly the same in every direction. This is clearly not the fact for a LASER beam. And also for the projection of a slide on a screen, it is clearly that the speed of light within the plane of the screen equals zero. Because the slide we observe does not move. While the projection beam itself moves towards the screen with the speed of light "c", the beam clearly remains focused and does not move within the plane, perpendicular to the direction of propagation.

There is no other conclusion than the conclusion that the Maxwell Equations are "wrong" or at least "not complete". The right equation(s) have to describe both possibilities. The possibility that the light moves in every direction with the exactly the same speed of light "c" like the light being emitted by the sun. And the possibility that the light moves only in one direction and equals zero in the directions perpendicular to the plane of propagation like the propagation of a LASER beam.

To find these new equation(s) we observe that the Maxwell equations are not in unification with Newton's theory of equilibrium of forces. The Maxwell Equations are not in unification with Newtons 3rd law "action = - reaction". Maxwell has not included the force densities with an electromagnetic field at all. To find this new equation, we have to introduce the force densities within an electromagnetic field. This has been presented in equation (3). The equilibrium of forces has been presented in equation (4). And the new equation found, has been presented in (5).

9

1.1 Impact of a New Theory on Modern Physics from the last 150 years

When we look at Modern Physics developed during the last 150 years, it is clearly to recognize fundamental important turning points during the last 150 years. A great and important turning point has been the fifth Solvay Conference in October 1927 where a fundamental debate took place between Niels Bohr (representing the instrumentalists) and Albert Einstein (representing the scientific realists). The instrumentalists wanted loser rules based on outcomes. The outcome of this fundamental debate between Niels Bohr and Albert Einstein was that the instrumentalists won the debate. Based on this outcome, the fundamental quantum physical "Copenhagen Interpretation", had been presented by Niels Bohr and and Werner Heisenberg and was generally accepted.

The fundamental question is: Would the development of the fundamental theories in Modern Physics had been different when Maxwell would have found a different set of equations for the electromagnetic field in his famous and well-known publication in 1865 "A Dynamical Theory of the Electromagnetic Field".

And indeed, the final and only conclusion will be that physics would have developed in a totally different way when Maxwell would have found a different set of equations, describing the electromagnetic field. When Maxwell would have found equation (5), presented in this book, the New Equation for the Electromagnetic field, the name "Quantum Physics" would probably never exist and the "Copenhagen Interpretation" would never have been formulated by Niels Bohr and Werner Heisenberg. The new name for Quantum Physics" would probably be something like "Quantum Electrodynamics" based on the fundamental properties of the electromagnetic field.

Because one of the possible outcomes (solutions) of equation (5) is the possibility of independently and separately existing "Electromagnetic Fields Configurations" confined by their own electromagnetic mutual interaction. Existing like elementary particles with their own distinctive frequencies and interacting like non deformable particles.

The underlaying conflict in the famous discussion between Niels Bohr and Albert Einstein at the fifth Solvay Conference in October 1927 was the impossibility to explain the matter waves proposed by Louis de Broglie within an electromagnetic concept.

The concept that matter behaves like a wave was proposed by Louis de Broglie, who introduced in 1924 the concept of "matter waves". Erwin Schrödinger published in 1926 the famous quantum mechanical Schrödinger wave equation. The Schrödinger wave equation formulated the "matter waves" proposed by "Louis de Broglie" within a mathematical concept.

Since then fundamental choices have been made, finally resulting in the "Copenhagen Interpretation" formally presented during the fifth Solvay Conference in October 1927 by Niels Bohr and Werner Heisenberg.

However, the well-known "Copenhagen Interpretation" has been a direct consequence of the wrong formulation of the Maxwell equations. Because the mathematical solutions for the Schrödinger wave equation" are complex solutions, describing spherical and elliptical waves.

It is impossible to find with the classical Maxwell equations the complex spherical and elliptical wave solutions which were found with the Schrödinger wave equation. And for that reason it was clearly impossible that matter waves could be electromagnetic waves. And a totally new kind of waves, indicated as "matter waves", had to be introduced.

11

In this book, equation (5) has been introduced, describing the electrodynamics of the electromagnetic field. And this equation (5) easily results in the spherical solutions, represented in equation (42), and the elliptical complex solutions. And for that reason electromagnetic waves are still a fundamental and important candidate for the quantum mechanical "matter waves".

Confined electromagnetic waves represent the property of inertia and like every kind of matter obey Newton's second law: $F = m.a$, presented in equation (36). This is a second argument to choose for confined electromagnetic waves as an important candidate for "matter waves"

Introducing two electromagnetic complex functions (5.1.1) and (5.1.2) and substituting both complex electromagnetic functions in the New Electromagnetic Equation 5.1 results in the quantum mechanical Dirac equation (5.1.7) which is a relativistic equivalent for the quantum mechanical Schrödinger wave equation. This is the third argument to choose for confined electromagnetic waves as an important candidate for "matter waves"

The 1927 Solvay Conference has been a fundamental and determining point in the history of Modern Physics. From that starting point the "quantum mechanical "Copenhagen Interpretation" has been the fundamental " Leitmotiv " in modern physics.

But this " Leitmotiv " has been based on the wrong equation. Because the Maxwell Equations are linear equations and do not have the fundamental mathematical complex solutions like the quantum mechanical Dirac equation and the Schrödinger wave equation. And for that fundamental reason "electromagnetic waves" could never be a candidate for the till then unknown "matter waves" or the "de Broglie" waves.

12

With the presentation of the new equation (5) a new set of fundamental mathematical complex solutions for confined electromagnetic waves has been offered. Solutions which match the mathematical solutions for the quantum mechanical Dirac Equation and the Schrödinger wave equation.

Introducing in this way a new " Leitmotiv " in Modern Physics based on the fundamental new concept that "Confined Electromagnetic Waves" with the mechanical properties of mass and momentum are a serious candidate for the "matter waves" or the "De Broglie waves". In this new concept the whole idea of "Probability Waves" combined with the "Wave-Particle Duality" will disappear.

1.2 Introduction into the Electromagnetic Theory to describe the propagation of a beam of light as well for the spherical emission of light (emitted by the sun) and for the one-directional emission of light (LASER beam).

Albert Einstein, Lorentz and Minkowski published in 1905 the Theory of Special Relativity and Einstein published in 1915 his field theory of general relativity based on a curved 4-dimensional space-time continuum to integrate the gravitational field and the electromagnetic field in one unified field. Since then the method of Einstein's unifying field theory has been developed by many others in more than 4 dimensions resulting finally in the well-known 10-dimensional and 11-dimensional "string theory".

 String theory is an outgrowth of S-matrix theory, a research program begun by Werner Heisenberg in 1943 (following John Archibald Wheeler's[3] 1937 introduction of the S-matrix), picked up and advocated by many prominent theorists starting in the late 1950's.

Theodor Franz Eduard Kaluza (1885-1954), was a German mathematician and physicist well-known for the Kaluza–Klein theory involving field equations in curved five-dimensional space. His idea that fundamental forces can be unified by introducing additional dimensions re-emerged much later in the "String Theory".

The original Kaluza-Klein theory was one of the first attempts to create an unified field theory i.e. the theory, which would unify all the forces under one fundamental law. It was published in 1921 by Theodor Kaluza and extended in 1926 by Oskar Klein. The basic idea of this theory was to postulate one extra compactified space dimension and introduce nothing but pure gravity in a new $(1 + 4)$-dimensional space-time. Klein suggested that the fifth dimension would be rolled up into a tiny, compact loop on the order of 10^{-35} [m]

To use simple notifications, the Einstein convention will be used. In the Einstein Convention the index always changes from 1 till 4. Others prefer the changing of the index from 0 till 3 but is has the same meaning.

This means the term: " $x_a y_a$ " has to be interpreted as:

$$x_a y_a = x_1 y_1 + x_2 y_2 + x_3 y_3 + x_4 y_4 \qquad (B.0$$

In Cartesian coordinates this means:

$$(1,2,3,4) \rightarrow (x, y, z, t) \text{ or } (0, 1, 2, 3) \rightarrow (t,x,y,z)$$

In Classical Electrodynamics, the Electromagnetic Field has been derived from the 4-dimensional Potential 4-Vector. The 4-dimensional Electromagnetic "Potential 4-vector", oriented in the classical complex 4-dimensional "Minkowski Space" will be defined by φ_a ("a" varying from 1 until 4) in which:

$$\varphi_a = \begin{pmatrix} i\,V/\,c \\ A_3 \\ A_2 \\ A_1 \end{pmatrix} \qquad (B.1)$$

15

In which V equals the scalar electric potential and $\overline{A}$ the 3-dimensional magnetic vector potential.

In which V equals the scalar electric potential and $\overline{A}$ the 3-dimensional magnetic vector potential. The Electric Field Intensity $\overline{E}$ equals:

$$\overline{E} = - \nabla V - \frac{\partial \overline{A}}{\partial t} \qquad \text{(B.2)}$$

The Magnetic Field Intensity $\overline{H}$ equals:

$$\overline{H} = \frac{1}{\mu_0} B = \frac{1}{\mu_0} \left(\nabla \times \overline{A} \right) \qquad \text{(B.3)}$$

The 4-dimensional Electromagnetic "Maxwell Tensor" has been defined by:

$$F_{ab} = \partial_b \varphi_a - \partial_a \varphi_b \qquad \text{(1)}$$

The 4-dimensional Electromagnetic "Energy Momentum Tensor" has been defined by:

$$T^{ab} = \frac{1}{\mu_0} \left[F_{ac} F^{cb} + \frac{1}{4} \delta_{ab} F_{cd} F^{cd} \right] \qquad \text{(2)}$$

The 4-dimensional divergence of the Energy Momentum Tensor equals the 4-dimensional Force Density 4-vector:

$$f^a = \partial_b T^{ab} \qquad \text{(3)}$$

The new theory has been based on the fundamental concept of Harmony in which all force densities in the Universe have been counter balanced by equal and opposite directed force densities resulting in a net force density equals zero and a final set of 4 Electromagnetic Equations.

$$f^a = \partial_b \, T^{ab} = 0 \tag{4}$$

In the absence of any Gravity, the force density f^a in the 3 directions of the 3 coordinates of the chosen 3-coordinate system follows from the (4-dimensional) Divergence of the (4-dimensional) Stress Energy Tensor [8,9,38] (3).

The Divergence of a Vector equals a Scalar. The Divergence of a Tensor equals a Vector. The 4-dimensional Divergence of the 4-dimensional Stress Energy Tensor (4) equals the 4-dimensional Force-density Vector.

The first 3 terms of the 4-dimensional Force-density vector equal the force densities in the corresponding 3 dimensions of the chosen Coordinate System. The 4th component equals the Electromagnetic Poynting's Theorhem (5.1) (Continuity Equation).

To calculate the equilibrium conditions to present the force densities in the Electromagnetic Field Configuration, the first 3 terms of the 4-dimensional Force-density vector are being used.

By re-arranging the first 3 terms of the (4-dimensional Divergence) of the (4-dimensional) Stress Energy Tensor (4) an equation for the 3-dimensional force density f^a within the Electromagnetic Field Configuration has been created. This Equation (5.2) represents the 3-dimensional force density f^a in a coordinate-free vector equation in the absence of any Gravity:

$$\left(x_4 \right) \qquad \nabla . \left(\bar{E} \times \bar{H} \right) + \frac{1}{2} \frac{\partial \left[\varepsilon_0 \left(\bar{E} . \bar{E} \right) + \mu_0 \left(\bar{H} . \bar{H} \right) \right]}{\partial t} = 0 \qquad (5.1)$$

$$(5.N)$$

$$\begin{pmatrix} x_3 \\ x_2 \\ x_1 \end{pmatrix} \quad \begin{aligned} & -\frac{1}{c^2} \frac{\partial \left(\bar{E} \times \bar{H} \right)}{\partial t} + \varepsilon_0 \, \bar{E} \left(\nabla . \bar{E} \right) - \varepsilon_0 \, \bar{E} \times \left(\nabla \times \bar{E} \right) + \\ & \\ & + \mu_0 \, \bar{H} \left(\nabla . \bar{H} \right) - \mu_0 \, \bar{H} \times \left(\nabla \times \bar{H} \right) = \bar{0} \end{aligned} \qquad (5.2)$$

According the fundamental requirement of a perfect equilibrium at any place at any time in any direction, the algebraic sum of all the different force densities $\vec{f}$ have to be equal zero for any physical possible electromagnetic field configuration (particles and fields).

It is fundamental to realize that 3 different kind of separate interactions of the force densities in (5.2) are being involved within this resulting equilibrium. Magnetic-Magnetic interaction (4th and 5th term in 5.2), Electric-Electric interaction (2nd and 3rd term in 5.2), Electric-Magnetic interaction and reverse (1st term in 5.2) which is time-dependent according the theory of special relativity.

An electromagnetic field which is in a perfect equilibrium with itself and its surrounding at any space and time in any direction, fulfills the necessary requirements for the physical possibility of the existence of this field. Under that condition Equation (4) transforms into the "Unified 4-Dimensional Hyperspace Equilibrium Equation" (5)

$$- \frac{1}{c^2} \frac{\partial (\overline{E} \times \overline{H})}{\partial t} + \varepsilon_0 \, \overline{E} \, (\nabla . \overline{E}) - \varepsilon_0 \overline{E} \times (\nabla \times \overline{E}) + \qquad (5)$$

$$+ \mu_0 \, \overline{H} \, (\nabla . \overline{H}) - \mu_0 \, \overline{H} \times (\nabla \times \overline{H}) = \overline{0}$$

To extend Field Equation (5) into an equilibrium within a multi-dimensional curved Space-Time continuum unifying different fields like gravity and electromagnetism, the transformation has been realized by the transformation of the resulting force-densities within the 4-Dimensional Space-Time continuum. The Unification of the Electromagnetic Fields with the Gravitational fields results in the Relativistic Gravitational Electro Magnetic Equilibrium (RGEE) equation

19

within a gravitational field $\vec{g}$ in the 3-dimensional (spatial) representation:

$$- \frac{1}{c^2} \frac{\partial (\overline{E} \times \overline{H})}{\partial t} + \varepsilon_0 \, \overline{E} \, (\nabla \cdot \overline{E}) - \varepsilon_0 \, \overline{E} \times (\nabla \times \overline{E}) + \mu_0 \, \overline{H} \, (\nabla \cdot \overline{H}) - \tag{5-a}$$

$$- \mu_0 \, \overline{H} \times (\nabla \times \overline{H}) - \tfrac{1}{2} \, \varepsilon_0^2 \mu_0 \left(\overline{E} \cdot \overline{E} \right) \overline{g} - \tfrac{1}{2} \varepsilon_0 \mu_0^2 \left(\overline{H} \cdot \overline{H} \right) \overline{g} = \overline{0}$$

It is fundamental to realize that Equation (5) is only a part of the 4-Dimensional Time-Space Continuum Equation.

The Divergence of the 4-Dimensional Stress-Energy Tensor (3) results in the 4-Dimensional Vector Equation. The first 3 terms of the vector Equation have been presented in (5). The 4th term presents the Continuum Equation. By introducing the complex field notations for the electric field and the magnetic field in Equation (39) the 4th term transforms into the well-known relativistic quantum mechanical Dirac Equation and at low velocities into the quantum mechanical Schrödinger Wave Equation.

The 4 Equations together (3 Equations for the separate space coordinates) and the Dirac/Schrödinger Equation describe the Unification in a perfect Equilibrium of the different Fields.

1.3 The 4th term in the Unified 4-Dimensional Hyperspace Equilibrium Equation

The 4-Dimensional Hyperspace Equilibrium Dirac Equation (5.1) in the absence of gravity equals:

$$(x_4) \quad \nabla . (\overline{E} \times \overline{H}) = -\frac{1}{2} \frac{\partial \left(\varepsilon_0 \left(\overline{E} . \overline{E} \right) + \mu_0 \left(\overline{H} . \overline{H} \right) \right)}{\partial t} \qquad (5.1)$$

$$(5)$$

$$\begin{pmatrix} x_3 \\ x_2 \\ x_1 \end{pmatrix} \quad -\frac{1}{c^2} \frac{\partial (\overline{E} \times \overline{H})}{\partial t} + \varepsilon_0 \overline{E} (\nabla . \overline{E}) - \varepsilon_0 \overline{E} \times (\nabla \times \overline{E}) +$$

$$+ \mu_0 \overline{H} (\nabla . \overline{H}) - \mu_0 \overline{H} \times (\nabla \times \overline{H}) = \overline{0} \qquad (5.2)$$

The Poynting Theorem (5.1) can be rewritten by introducing the vector functions $\overline{\phi}$ and the complex conjugated vector function $\overline{\phi}*$ in which:

$$\overline{\phi} = \frac{1}{\sqrt{2\mu}} \left(\overline{B} + i \frac{\overline{E}}{c} \right) \qquad (5.1.$$

$\overline{B}$ equals the magnetic induction, $\overline{E}$ the electric field intensity and c the speed of light. The complex conjugated vector function equals:

$$\overline{\phi}* = \frac{1}{\sqrt{2\mu}} \left(\overline{B} - i \frac{\overline{E}}{c} \right) \qquad (5.1.$$

The dot product equals the electromagnetic energy density w:

$$\overline{\phi} . \overline{\phi}* = \frac{1}{2\mu} \left(\overline{B} + i \frac{\overline{E}}{c} \right) . \left(\overline{B} - i \frac{\overline{E}}{c} \right) = \frac{1}{2} \mu H^2 + \frac{1}{2} \varepsilon E^2 = w \qquad (5.1$$

The cross product is proportional to the Poynting vector (ref. 29, page 202, equation 15).

$$\bar{\phi} \times \bar{\phi}^* = \frac{1}{2\mu}\left(\bar{B} + i\,\frac{\bar{E}}{c}\right) \times \left(\bar{B} - i\,\frac{\bar{E}}{c}\right) = i\sqrt{\varepsilon\mu}\;\bar{E} \times \bar{H} = i\sqrt{\varepsilon\mu}\;\bar{S} \qquad (5.1.4)$$

Substituting (5.P.3) and (5.P.4) in Equation (5) results in The 4-Dimensional Hyperspace Equilibrium Equation:

$$(x_4) \quad -\frac{i}{\sqrt{\varepsilon_0\,\mu_0}}\nabla \cdot (\bar{\phi} \times \bar{\phi}) = -\frac{\partial \bar{\phi} \cdot \bar{\phi}^*}{\partial t} \qquad\qquad (5.1.5)$$

$$(5.5)$$

$$\begin{pmatrix} x_3 \\ x_2 \\ x_1 \end{pmatrix} \quad \begin{aligned} &-\frac{1}{c^2}\frac{\partial\,(\bar{E} \times \bar{H})}{\partial t} + \varepsilon_0\,\bar{E}\,(\nabla \cdot \bar{E}) - \varepsilon_0\,\bar{E} \times (\nabla \times \bar{E}) + \\[1em] &+ \mu_0\,\bar{H}\,(\nabla \cdot \bar{H}) - \mu_0\,\bar{H} \times (\nabla \times \bar{H}) = \bar{0} \end{aligned} \qquad (5.2.5)$$

To transform the electromagnetic vector wave function $\bar{\phi}$ into a scalar (spinor or one-dimensional matrix representation), the Pauli spin matrices σ and the following matrices (Ref. 29 page 213, equation 99) are introduced:

$$\bar{\alpha} = \begin{bmatrix} 0 & \sigma \\ \sigma & 0 \end{bmatrix} \quad \text{and} \quad \bar{\beta} = \begin{bmatrix} \delta_{ab} & 0 \\ 0 & -\delta_{ab} \end{bmatrix} \qquad (5.1.6)$$

22

Then equation (5) can be written as the 4-Dimensional Hyperspace Equilibrium Dirac Equation:

$$(x_4) \quad \left(\frac{i\,m\,c}{h} \overline{\beta} + \overline{\alpha} \cdot \nabla \right) \psi \; = \; -\frac{1}{c} \frac{\partial \psi}{\partial t} \tag{5.1.7}$$

$$\left(5.7 \right.$$

$$\begin{pmatrix} x_3 \\ x_2 \\ x_1 \end{pmatrix} \quad -\frac{1}{c^2} \frac{\partial\,(\overline{E} \times \overline{H})}{\partial t} \; + \; \varepsilon_0 \, \overline{E} \; (\nabla \cdot \overline{E}) - \varepsilon_0 \, \overline{E} \times (\nabla \times \overline{E}) \; + \tag{5.2.7}$$

$$+ \; \mu_0 \, \overline{H} \; (\nabla \cdot \overline{H}) - \mu_0 \, \overline{H} \times (\nabla \times \overline{H}) \; = \overline{0}$$

The fourth term (x_4) equals the relativistic Dirac equation (5.1.7) which equals equation (102) page 213 in Ref.29.

1.3   The Impact of Gravity on Light

We consider a beam of light approaching a strong gravitational field. (E.g. a Black Hole). According the first term in equation (5.2) the beam of light will follow a circular orbit around the Black Hole. The required Equilibrium will exist at the radius where the centrifugal electromagnetic inertia forces will be equal and opposite directed to the centripetal oriented gravitational forces on the electromagnetic mass.

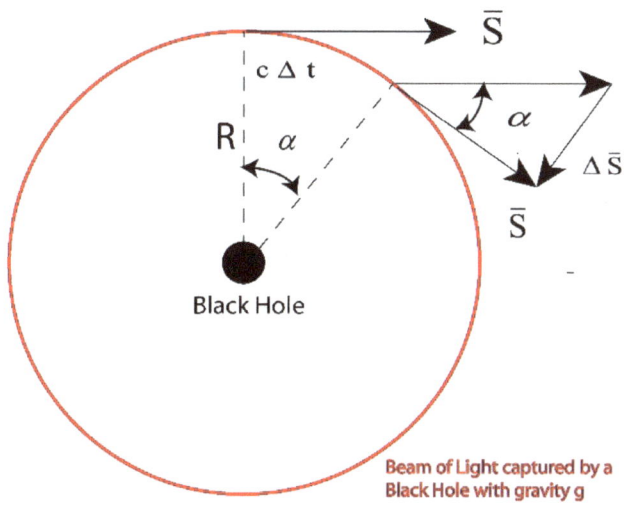

Beam of Light captured by a
Black Hole with gravity g

The whole Universe is in a perfect Equilibrium. This fundamental property of Equilibrium has been extended to a 4-dimensional Hyperspace Continuum in which a perfect equilibrium persists in any of the 4 coordinate directions.

The requirement of a 4-dimensional Equilibrium results in the outcome that the Dirac Equation is only one equation in a set of 4 equations. And that the Dirac Equation originates from an electromagnetic equation in the time-energy domain. This new 4-Dimensional Hyperspace Equilibrium Theory opens a new door to an unexplored field of mathematical and physical challenges. This theory is a new approach in physics based on a 4-Dimensional Hyperspace Equilibrium resulting in the 4-dimensional Dirac Equation which represents the "Unification of (quantum mechanical) De Broglie Waves and Electromagnetic Waves". Solving these 4 simultaneous equations offers the possibility to find answers to the fundamental questions in physics within a quantum mechanical 4-Dimensional Frame-Work. Every Physical Possible Electro-Magnetic Field Configuration of Confinement has to be a solution of this fundamental 4-Dimensional Relativistic Dirac Equation.

Energy-Time Domain

$$(x_4) \qquad \left(\frac{i\,m\,c}{h} \, \overline{\beta} + \overline{\alpha} \cdot \nabla \right) \psi = - \frac{1}{c} \, \frac{\partial \psi}{\partial t}$$

3-Dimensional Space Domain

$$\begin{pmatrix} x_3 \\ \\ x_2 \\ \\ x_1 \end{pmatrix}$$

$$\overset{\text{B-1}}{- \frac{1}{c^2} \frac{\partial\,(\overline{E} \times \overline{H})}{\partial t}} + \overset{\text{B-2}}{\varepsilon_0 \, \overline{E} \, (\nabla \cdot \overline{E})} - \overset{\text{B-3}}{\varepsilon_0 \, \overline{E} \times (\nabla \times \overline{E})} +$$

$$\overset{\text{B-4}}{+ \mu_0 \, \overline{H} \, (\nabla \cdot \overline{H})} - \overset{\text{B-5}}{\mu_0 \, \overline{H} \times (\nabla \times \overline{H})} = \overline{0}$$

The Equation in the Time-Energy Domain (x4) equals the well-known Quantum Mechanical Dirac Equation. The 3 Equations in the 3 spatial directions (coordinates x1, x2, x3) describe the Electromagentic Field. The term B-1 controls the speed of light. The terms B-2 and B-3 control the confinement of Light for the Electric Radiation Pressure and the terms B-4 and B-5 control the confinement of Light for the Magnetic Radiation Pressure.

As a demonstration of the power of the 4-dimensional Relativistic Dirac Equation, we observe a very basic and simple polarized laser beam with a Gaussian Intensity Division. The laser beam propagates towards the z-direction. The Electric Field "E" is oriented along the x-axis and the Magnetic Field "H" is oriented along the y-axis.

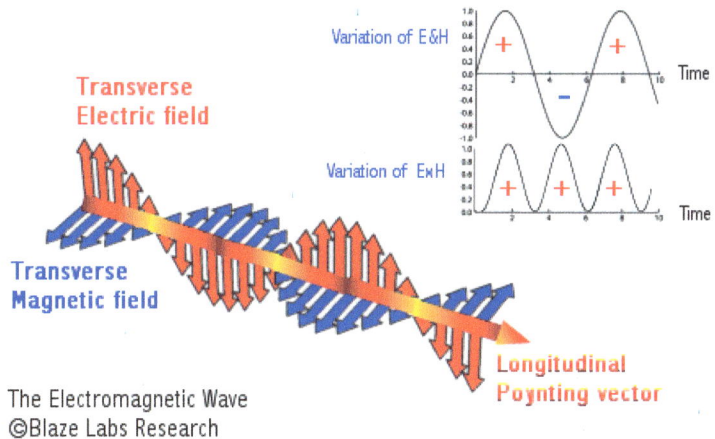

The Electromagnetic Wave
©Blaze Labs Research

The laser beam propagates towards the positive z-direction:

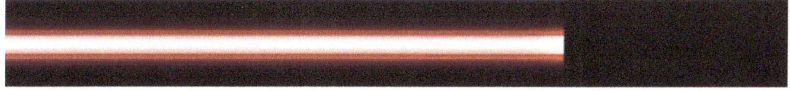

The laser beam propagates towards the positive z-direction with the speed of light. That is only possible because exactly at the speed of light there will exist a "Perfect Equilibrium" between the Electromagnetic Radiation Pressure towards the positive z-direction and the inertia term of the Electromagnetic Radiation Energy (electromagnetic mass) which has been represented by the time derivative of the "Poynting Vector" in term B-1 and is oriented along the negative z-direction.

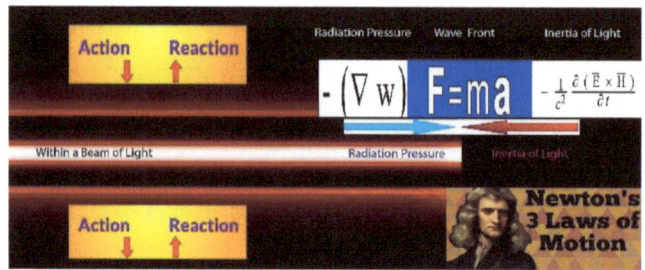

In the Unified 4-dimensional Equilibrium always a"Perfect Equilibrium" will exist between the Electromagnetic Radiation Pressure towards the positive z-direction and the inertia term of the Electromagnetic Radiation Energy (electromagnetic mass) which has been represented by the time derivative of the "Poynting Vector" in term B-1 and is oriented along the negative z-direction.

The inward oriented Electromagnetic Force Density in the z-direction for every arbitrary function f(x,y,z) propagating with the speed of light $c = 1 / \sqrt{\varepsilon_0 \, \mu_0}$:

$$f_z = - \varepsilon_0 \, \mu_0 \, \frac{\partial \, \overline{E} \, \times \, \overline{H}}{\partial \, t} = - 2 \, \varepsilon_0^{3/2} \, \sqrt{\mu_0} \, f \, (x \, , y)^2 \, g\left(t - z \, \sqrt{\varepsilon_0 \, \mu_0}\right) g'\left(t - z \, \sqrt{\varepsilon_0 \, \mu_0}\right)$$

Compensates the outward oriented Electromagnetic Radiation Pressure in the z-direction for every arbitrary function f(x,y,z) propagating with the speed of light $c = 1 / \sqrt{\varepsilon_0 \, \mu_0}$:

$$f_z = - \left(\nabla \, w \right)_z = 2 \, \varepsilon_0^{3/2} \, \sqrt{\mu_0} \, f \, (x \, , y)^2 \, g\left(t - z \, \sqrt{\varepsilon_0 \, \mu_0}\right) g'\left(t - z \, \sqrt{\varepsilon_0 \, \mu_0}\right)$$

The electromagnetic "Radiation Pressure" in the positive z-direction (direction of propagation) has been determined by the electromagnetic "Energy Density". The electromagnetic "Inertia Force" has been determined by the electromagnetic "Mass (Energy Density)" and the "Propagation Speed". There is only "One Exact Single Velocity" at which "Velocity" the electromagnetic "Radiation Pressure" in the in the positive z-direction perfectly counterbalances the opposite directed (negative z-direction) "Force of Inertia". That single speed has been called the "Velocity of Perfect Equilibrium" and has been called in general the "Speed of Light" and is independent of the frequency and the intensity of the electromagnetic beam of light.

27

$$\text{Velocity of Perfect Equilibrium } = c = \frac{1}{\sqrt{\varepsilon_0\,\mu_0}} = 299{,}792 \text{ [km /s]}$$

When a laser beam (beam of light) passes a gravitational field with acceleration "g" in the (x,y) direction, the radiation pressures within de the beam of light in the (x,y) plane will shift due to gravitational forces. Because according Einstein's $E = mc^2$, the electromagnetic energy of the beam of light has electromagnetic mass, which will be attracted by the gravitational field. The shift in the radiation pressures in the x-y plane due to gravitational-electromagnetic interaction can only be compensated by the inertia forces within the (x,y) plane due to a circular orbit of the beam of light with the origin of the gravitational field as the center.

$$\text{Tan}(\alpha) = \frac{c\,\Delta t}{R} = \frac{\Delta S}{S}$$

Newton: F (Force) = m (mass) a (accelleration)

in which f (force density [N/m^2]) and ρ (specific mass [kg/m^3])

$$f_{INERTIA} = \frac{1}{c^2}\frac{\Delta S}{\Delta t} = \frac{S}{R\,c} = \rho\,g$$

Einstein (E = m c^2)

$$\frac{1}{c^2}\frac{\Delta S}{\Delta t} = \frac{S}{R\,c} = \rho\,g = \frac{w}{c^2}\,g = \frac{\varepsilon\,E^2 + \mu\,H^2}{2\,c^2}\,g$$

$$\frac{E}{H} = \sqrt{\frac{\mu}{\varepsilon}}$$

$$R = \frac{S}{\rho\,c\,g} = \frac{E^2\,c\sqrt{\frac{\varepsilon}{\mu}}}{g\,\varepsilon\,E^2} = \frac{1}{\varepsilon\,\mu\,g} \approx \frac{9\;10^{16}}{g}$$

The perfect equilibrium, where the inertia forces due to the circular orbit of the beam of light are in a perfect balance with the attractive gravitational forces, exists at one defined radius "R" of the beam of light (laser beam), independent of the intensity of the beam of light and independent of the frequency of the beam of light. Only the acceleration "g" of the gravitational field determines the radius of equilibrium "R"

$$R \approx \frac{9\ 10^{16}}{g}$$

In which R the radius of the beam of light, and g the
accelleration of the gravitational field of the Black Hole

The x-y plane is oriented perpendicular on the z-direction.
The speed of light towards the positive z-direction equals the
speed of light (the constant "c = 300.000 km/s"). But the
speed of light in the x-y plane has to be exactly zero. Else the
diameter of the laser beam would become larger and larger
during the propagation along the positive z-direction. This is
only possible becasue the Electromagnetic confining forces
B-2,B-3,B-4 and B-5 compensate exactly the outward
oriented radiation pressure towards the x-dirention and the y-
direction.

The Electric Radiation Pressure has been compensated by the
Coulomb Force Densities within the Laser Beam

Coulomb's Law

$\overline{F} = \overline{E}\,Q$ $(F \triangleq$ Force and $Q \triangleq$ Electric Charge)

$\overline{f} = \overline{E}\,\rho$ $(f \triangleq$ force density and $\rho \triangleq$ charge density)

$\overline{f} = \overline{E}\,\nabla . \overline{D}$ $(D \triangleq$ Dielectric Displacement)

$\overline{f} = \varepsilon_0\,\overline{E}\,(\nabla . \overline{E})$ $(\varepsilon_0 \triangleq$ permittivity)

$$(f_4) \quad \left(\frac{i\,m\,c}{h}\,\overline{\beta} + \overline{\alpha}.\nabla\right)\psi = -\frac{1}{c}\frac{\partial\psi}{\partial t}$$

$$\begin{pmatrix} f_3 \\ f_2 \\ f_1 \end{pmatrix} \quad -\frac{1}{c^2}\frac{\partial\,(\overline{E}\times\overline{H})}{\partial t} + \varepsilon_0\,\overline{E}\,(\nabla.\overline{E}) - \varepsilon_0\,\overline{E}\times(\nabla\times\overline{E}) +$$

$$+ \mu_0\,\overline{H}\,(\nabla.\overline{H}) - \mu_0\,\overline{H}\times(\nabla\times\overline{H}) = \overline{0}$$

The Electric Radiation Pressure along the x-axis has been
compensated by the term B-2:

The inward oriented Electric Force Density in the x-direction for every arbitrary function f(x,y,z):

$$f_x = \varepsilon_0 \, E_x \, \frac{\partial \, E_x}{\partial \, x} = \varepsilon_0 \; f\left(x,y\right) \, g\left(t - z \sqrt{\varepsilon_0 \, \mu_0}\right)^2 f^{(1,0)}\left(x,y\right)$$

Compensates the outward oriented Electric Radiation Pressure in the x-direction for every arbitrary function f(x,y,z):

$$f_x = -\left(\nabla \, w\right)_x = -\varepsilon_0 \; f\left(x,y\right) \, g\left(t - z \sqrt{\varepsilon_0 \, \mu_0}\right)^2 f^{(1,0)}\left(x,y\right)$$

The Electric Radiation Pressure along the y-axis has been compensated by the term B-3:

The inward oriented Electric Force Density in the y-direction for every arbitrary function f(x,y,z):

$$f_y = -\varepsilon_0\left(\overline{E} \times\left(\nabla\times\overline{E}\right)\right)_y = -\varepsilon_0 \; f\left(x,y\right) \, g\left(t - z \sqrt{\varepsilon_0 \, \mu_0}\right)^2 f^{(0,1)}\left(x,y\right)$$

Compensates the outward oriented Electric Radiation Pressure in the y-direction for every arbitrary function f(x,y,z):

$$f_y = -\left(\nabla \, w\right)_y = \varepsilon_0 \; f\left(x,y\right) \, g\left(t - z \sqrt{\varepsilon_0 \, \mu_0}\right)^2 f^{(0,1)}\left(x,y\right)$$

The Magnetic Radiation Pressure along the y-axis has been compensated by the term B-4:

The inward oriented Magnetic Force Density in the y-direction for every arbitrary function f(x,y,z):

$$f_y = \mu_0 \, H_y \, \frac{\partial \, H_y}{\partial \, y} = \varepsilon_0 \; f\left(x,y\right) \, g\left(t - z \sqrt{\varepsilon_0 \, \mu_0}\right)^2 f^{(0,1)}\left(x,y\right)$$

Compensates the outward oriented Magnetic Radiation Pressure in the y-direction for every arbitrary function f(x,y,z):

$$f_y = -\left(\nabla \, w\right)_y = -\varepsilon_0 \; f\left(x,y\right) \, g\left(t - z \sqrt{\varepsilon_0 \, \mu_0}\right)^2 f^{(0,1)}\left(x,y\right)$$

The Magnetic Radiation Pressure along the x-axis has been compensated by the term B-5:

The inward oriented Magnetic Force Density in the x-direction for every arbitrary function f(x,y,z):

$$f_x = -\varepsilon_0 \left(\overline{H} \times (\nabla \times \overline{H}) \right)_x = -\varepsilon_0 \ f(x,y) \ g\left(t - z\sqrt{\varepsilon_0\,\mu_0}\right)^2 f^{(1,0)}(x,y)$$

Compensates the outward oriented Magnetic Radiation Pressure in the x-direction for every arbitrary function f(x,y,z):

$$f_x = -(\nabla w)_x = \varepsilon_0 \ f(x,y) \ g\left(t - z\sqrt{\varepsilon_0\,\mu_0}\right)^2 f^{(1,0)}(x,y)$$

2. 1 EM Radiation within a Cartesian Coordinate System in the absence of gravity

The required Electromagnetic Field Configuration for a perfect Equilibrium in Space and Time follows from the dynamic equilibrium equation (5) and equals in Cartesian Coordinates $\{x, y, z, t\}$ for the Electric Field Components e (x, y, z, t):

$$
\begin{pmatrix} e_x \\ e_y \\ e_z \end{pmatrix} = \begin{pmatrix} f(x, y)\, g\left(t - \left(\dfrac{K_1}{z} + 1 \right) z \sqrt{\varepsilon_0\ \mu_0} \right) \\ 0 \\ 0 \end{pmatrix} \tag{6}
$$

The required Electromagnetic Field Configuration for a perfect Equilibrium in Space and Time follows from the dynamic equilibrium equation (5) and equals in Cartesian Coordinates $\{x, y, z, t\}$ for the Magnetic Field Components m (x, y, z, t):

$$
\begin{pmatrix} m_x \\ m_y \\ m_z \end{pmatrix} = \sqrt{\dfrac{\varepsilon_0}{\mu_0}} \begin{pmatrix} 0 \\ f(x, y) g\left(t - \left(\dfrac{K_1}{z} + 1 \right) z \sqrt{\varepsilon_0\ \mu_0} \right) \\ 0 \end{pmatrix} \tag{7}
$$

In which K_1 is an arbitrary constant. For the divergence-free function $f(x, y) = 1$, the solutions (6) and (7) are also the solutions for the known Maxwell Equations. For the non-divergence-free functions $f(x, y)$, the solutions (6) and (7) are not solutions for the Maxwell Equations, which requires divergence-free electromagnetic waves, propagating with the speed of light $c = 1/\sqrt{\varepsilon_0 \mu_0}$, in the absence of any matter.

But they are solutions of the Dynamic Equilibrium Equation (5) and clearly do exist in physics. Comparable with the projection of a slide with a beamer on a flat screen in the z-direction. In which the slide has an arbitrary intensity-division $f(x, y)$. The information $f(x, y)$ on the slide propagates with the speed of light $c = 1/\sqrt{\varepsilon_0 \mu_0}$ towards the screen in the z-direction in this example.

2. 1.1 Laser Beam with a Gaussian division in the x-y plane within a Cartesian Coordinate System in the absence of gravity

The required <u>Electromagnetic Field Configuration</u> for a perfect Equilibrium in Space and Time follows from the dynamic equilibrium equation (5).

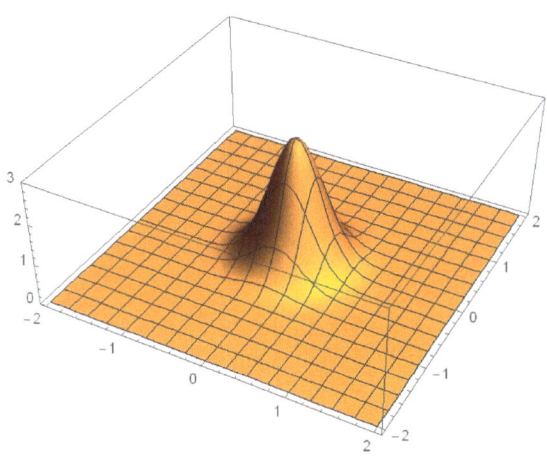

Figure 1. Electromagnetic Field Intensity with a Gaussian division

The required <u>Electromagnetic Field Configuration</u> equals in Cartesian Coordinates $\{x, y, z, t\}$ for the Electric Field Components $e(x, y, z, t)$:

$$
\begin{pmatrix} e_x \\ e_y \\ e_z \end{pmatrix} = \begin{pmatrix} K_1\, e^{-K_2\left(x^2+y^2\right)}\, \mathrm{Sin}\left(\omega\left(t - z\sqrt{\varepsilon_0\,\mu_0}\right)\right) \\ 0 \\ 0 \end{pmatrix}
$$

(

The required Electromagnetic Field Configuration for a perfect Equilibrium in Space and Time follows from the dynamic equilibrium equation (5) and equals in Cartesian Coordinates $\{x, y, z, t\}$ for the Magnetic Field Components m (x, y, z, t):

$$
\begin{pmatrix} m_x \\ m_y \\ m_z \end{pmatrix} = \sqrt{\frac{\varepsilon_0}{\mu_0}} \begin{pmatrix} 0 \\ K_1\, e^{-K_2\left(x^2+y^2\right)}\, \mathrm{Sin}\left(\omega\left(t - z\sqrt{\varepsilon_0\,\mu_0}\right)\right) \\ 0 \end{pmatrix} \quad (7.
$$

In which K_1 is an arbitrary constant. The Gaussian division is *not a solution* of the Maxwell Equations, because the divergence of the Electric field intensity *does not equal zero*

$$
\nabla \cdot \overline{E} = -2\, e^{-K_2(x^2+y^2)}\, x\, K_1\, K_2\, \mathrm{Sin}\left(\left(t - z\sqrt{\varepsilon_0\,\mu_0}\right)\omega\right) \quad (6.1
$$

And the divergence of the Magnetic field intensity *does not equal zero*.

35

$$\nabla \cdot \overline{H} = - \sqrt{\frac{\varepsilon_0}{\mu_0}} \left(2\, e^{-K_2(x^2+y^2)}\, y\, K_1\, K_2\, \mathrm{Sin}\left(\left(t\, -\, z\, \sqrt{\varepsilon_0\, \mu_0} \right) \omega \right) \right) \qquad (7.1)$$

For the non-divergence-free functions

$$f(x,y) = K_1\, e^{-K_2\left(x^2+y^2\right)}\, \mathrm{Sin}\left(\omega\left(t - z\sqrt{\varepsilon_0\, \mu_0} \right) \right),\ \text{the}$$

solutions (6) and (7) are not solutions for the Maxwell Equations, which requires divergence-free electromagnetic waves, propagating with the speed of light $c = 1/\sqrt{\varepsilon_0\, \mu_0}$, in the absence of any matter. But they are solutions of the Dynamic Equilibrium Equation (5) and clearly do exist in physics. Comparable with the projection of a slide with a beamer on a flat screen in the z-direction. In which the slide has an arbitrary intensity-division $f(x, y)$ and the intensity is not Divergence Free. The information $f(x, y)$ on the slide propagates with the speed of light $c = 1/\sqrt{\varepsilon_0\, \mu_0}$ towards the screen in the z-direction in this example and is a real physical possibility and is not a solution of the Maxwell equations but is a solution of the Dynamic Equilibrium Equation (5) and clearly do exist in physics.

2.2 EM Radiation within a Cartesian Coordinate System under the influence of a Longitudinal Gravitational Field g

The required Electromagnetic Field Configuration for a perfect Equilibrium in Space and Time for a Longitudinal Gravitational Field (The Light propagates in the same z-direction as the z-direction of the Gravitational Field) follows from the Dynamic Equilibrium Equation (5) and equals in Cartesian Coordinates $\{x, y, z, t\}$ for a gravitational field "g" for the Electric Field Components $e(x, y, z, t)$:

$$
\begin{pmatrix} e_x \\ e_y \\ e_z \end{pmatrix} = \begin{pmatrix} e^{-\frac{1}{2} g\, \varepsilon_0\, \mu_0\, z} & f(x, y) g\left(t - z\sqrt{\varepsilon_0\ \mu_0}\right) \\ 0 \\ 0 \end{pmatrix}
$$

(6–

The required Electromagnetic Field Configuration for a perfect Equilibrium in Space and Time for a Longitudinal Gravitational Field (The Light propagates in the same z-direction as the z-direction of the Gravitational Field) follows from the Dynamic Equilibrium Equation (5-a) and equals in Cartesian Coordinates $\{x, y, z, t\}$ for a gravitational field "g" for the Magnetic Field Components $m(x, y, z, t)$:

$$\begin{pmatrix} m_x \\ m_y \\ m_z \end{pmatrix} = \sqrt{\frac{\varepsilon_0}{\mu_0}} \begin{pmatrix} 0 \\ e^{-\frac{1}{2} g\,\varepsilon_0\,\mu_0\,z}\ f(x,y)\,g\left(t - z\sqrt{\varepsilon_0\,\mu_0}\right) \\ 0 \end{pmatrix} \qquad (7-a)$$

Equation (6-a) and (7-a) are solutions of (5-a) under the influence of a Longitudinal Gravitational field with field intensity "g" that acts along the z-direction while the electromagnetic wave is also propagating in the z-direction.

The electromagnetic wave is propagating with the unaltered speed of light $c = 1/\sqrt{\varepsilon_0\,\mu_0}$, independently of the strength g of the gravitational field in the z-direction. However, the amplitude of the electromagnetic wave becomes dependently of the gravitational intensity "g" and the distance "z" and changes along the z-axis with the electromagnetic-gravitational interaction term $e^{-\frac{1}{2} g\,\varepsilon_0\,\mu_0\,z}$.

In this example is chosen for e.g. a laser beam positioned vertically on the ground on earth , shining vertically against the gravitational field "g" of the earth. Because the laser beam presents electromagnetic energy, the beam has electromagnetic mass. The potential energy of the electromagnetic mass is increasing while the laser light is propagating upwards, against the direction of the gravitational field. Because of the law of conservation of Energy, the electromagnetic energy is decreasing over a distance "z" proportional with the same amount $e^{-g\,\varepsilon_0\,\mu_0\,z}$ as the potential energy of the electromagnetic mass is increasing.

2. 3 The Real Light Intensity of the Sun, measured in our Solar System, including Electromagnetic Gravitational Conversion (EMGC)

When a beam of light leaves the surface of the sun, the light will travel in the radial direction of the radial gravitational field caused by the sun. The required Electromagnetic Field Configuration for a perfect Equilibrium in Space and Time for a Radial Gravitational Field (The Light propagates in the same radial-direction as the radial-direction of the Gravitational Field) follows from the Dynamic Equilibrium Equation (5) and equals in Spherical Coordinates $\{r,\theta,\varphi,t\}$ for a gravitational field "g(r)" for the Electric Field Components $e(r,\theta,\varphi,t)$:

$$
\begin{pmatrix} e_r \\ e_\theta \\ e_\varphi \end{pmatrix} = \begin{pmatrix} \dfrac{1}{r}\ e^{\frac{G\,m_1\,\varepsilon_0\,\mu_0}{2\,r}} & K1 & g\!\left(t - r\ \sqrt{\varepsilon_0\,\mu_0}\right) \\ 0 \\ 0 \end{pmatrix}
$$

(6 – b

The required Electromagnetic Field Configuration for a perfect Equilibrium in Space and Time for a Radial Gravitational Field (The Light propagates in the same radial-direction as the radial-direction of the Gravitational Field) follows from the Dynamic Equilibrium Equation (5) and equals in Spherical Coordinates $\{r,\theta,\varphi,t\}$ for a gravitational field "g(r)" for the Magnetic Field Components $m(r,\theta,\varphi,t)$:

$$
\begin{pmatrix} m_x \\ m_y \\ m_z \end{pmatrix} = \sqrt{\frac{\varepsilon_0}{\mu_0}} \begin{pmatrix} 0 \\ \dfrac{1}{r} \; e^{\frac{G\,m_1\,\varepsilon_0\,\mu_0}{2\,r}} \quad K1 \;\; g\left(t - r\,\sqrt{\varepsilon_0\,\mu_0}\,\right) \\ 0 \end{pmatrix} \quad (7-b)
$$

Equation (6-b) and (7-b) are solutions of (5-a) under the influence of a Radial Gravitational field with field a gravitational field intensity "g(r)" that acts along the radial-direction while the electromagnetic wave is also propagating in the radial-direction.

When a light beam leaves the surface of the sun, the intensity will decrease according (6-b). At earth, the measured intensity will be according (5-a) and (6-b):

$$
I = \frac{I_0 \; e^{\frac{G\,m_1\,\varepsilon_0\,\mu_0}{r}}}{4\,\pi\,r^2} \qquad (6\text{-c})
$$

A beam of light represents an amount of electromagnetic energy. Which equals an amount of electromagnetic mass. This amount of electromagnetic mass is moving with the speed of light in the opposite direction of a (radial) gravitational field and gains potential energy. Because the law of conservation of energy , a part of the electromagnetic energy of the light beam has to be converted into potential energy according equation (6-c).

40

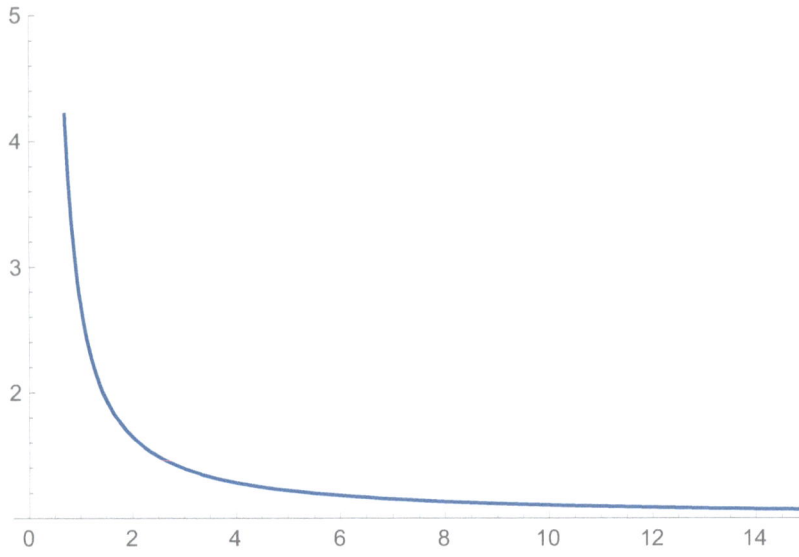

Figure 2. Electromagnetic Gravitational Conversion
Term $\quad e^{\frac{G\,m_1\,\varepsilon_0\,\mu_0}{r}}$

For a radius of the sun equals 695,508 [km] and a distance from the sun to the earth of 149,600,000 [km], the Electromagnetic Gravitational Conversion (EMGC) term equals:

$$C_{EMGC} = \frac{e^{\frac{G\,m_1\,\varepsilon_0\,\mu_0}{r_1}}}{e^{\frac{G\,m_1\,\varepsilon_0\,\mu_0}{r2}}} = \frac{e^{\frac{1}{r_1}}}{e^{\frac{1}{r2}}} = 4.1877534 \qquad (6\text{-}$$

This means that the **real intensity** of the light at the surface of the sun is about 4 times higher than the intensity which would have been calculated in a classical way from the sunlight intensity measured on earth, due to Electromagnetic Gravitational Conversion.

41

Equations (6-b) and (7-b) are solutions of (5-a) under the influence of a Radial Gravitational field with field intensity "g(r)" that acts along the radial-direction while the electromagnetic wave is also propagating in the same radial-direction. The electromagnetic wave is propagating with the unaltered speed of light $c = 1/\sqrt{\varepsilon_0 \mu_0}$, independently of the strength g(r) of the gravitational field in the radial-direction. However, the amplitude of the electromagnetic wave becomes dependently of the gravitational intensity "g(r)" and the distance "r" and changes along the radial direction due to the Electromagnetic-Gravitational Conversion term

$$C_{EMGC} = e^{\frac{G m_1 \varepsilon_0 \mu_0}{r}} .$$

Because of the law of conservation of Energy, the electromagnetic energy of the light emitted by the sun is decreasing over a distance "r" proportional with the same amount $EmGC = e^{\frac{G m_1 \varepsilon_0 \mu_0}{r}}$ as the potential energy of the electromagnetic mass of the light emitted by the sun is increasing.

2.4 The Boundaries of our Universe

When a mass , for example a ball, has been thrown into the air, the mass moves against the direction of the gravitational field of the earth and kinetic energy has been converted into potential energy. When the ball has reached the maximum height, the ball falls back towards the earth and the potential energy has been converted again into kinetic energy. The total sum of kinetic energy and potential energy remains constant.

The Cosmic Microwave Background Radiation (CMBR) is light (electromagnetic radiation) and represents an amount of electromagnetic energy. Which equals an amount of electromagnetic mass. This amount of electromagnetic mass is moving with the speed of light in the opposite direction of a (radial) gravitational field and gains potential energy. Because the law of conservation of energy , a part of the electromagnetic energy of the light beam has to be converted into potential energy according equation (6-c).

In a comparable way the Cosmic Microwave Background Radiation (CMBR) (light, electromagnetic) radiation moves away from the center of our Universe, gaining Potential Energy by moving against the (Radial) Gravitational field of our Universe. Until the Intensity of the light vanishes into zero at the boundaries of our Universe and has the potency to fall back towards its origin gaining an non-imaginable amount of light intensity by the conversion of potential energy into electromagnetic energy.

In this model of the Universe, the electromagnetic radiation within the universe, is locked up in its own gravitational

43

field. And the whole Universe can be considered to be a gigantic Black Hole with a diameter over 5 billion lightyears.

Depending on the the electromagnetic stability at the boundaries of our Universe a periodic model of our Universe is possible like the ball being thrown into the air and falling back towards the location where it came from, or the universe finally disappearing into the infinite space.

These possibilities depend on the radial dependent electromagnetic intensity function of the Cosmic Microwave Background Radiation (CMBR) Table 1 .

2.5 The Origin of Dark Matter

The Cosmic Microwave Background Radiation (CMBR) is the dominant radiation field in the Universe, and one of the most powerful cosmological tools that has yet been found, 25 years after its discovery by Penzias & Wilson (1965) .

Within a few years of the discovery of the CMBR, it was established the radiation field is close to isotropic, with a spectrum characterized by a single temperature, $T_{rad} \approx 2.7$ K. The specific intensity of the radiation is therefore close to:

$$I_f = \frac{2\,h\,f^3}{c^2}\ \left(e^{hf/k_B T_{rad}} - 1 \right)^{-1} \tag{6-e}$$

which corresponds to a peak brightness $I_{max} \sim 3.7$ x 10^{-18} W m^{-2} Hz^{-1} sr^{-1} at f$_{max}$ ~ 160 GHz and an energy density $i_f \sim 4$ x 10^{-14} J m^{-3}, which can also be expressed as a mass density ϱ_{em} ~ 5 x 10^{-31} kg m^{-3}.

When from the earth an Electromagnetic Mass Density has been measured which equals $\rho_{em} = 5 \times 10^{-36}$ kg/m^3, the Total Mass of the Universe can be roughly calculated without and with regarding the effect of Electromagnetic Gravitational Conversion.

At the Origin of the Universe, at the start of the Big Bang, a large amount of electromagnetic radiation has been blown into the universe, which has been measured on earth as the well-known CMBR. The radiation, traveling for billions of years against the gravitational field of the origin of the universe and has gained during that time an enormous amount of potential energy. Because of the conversation of energy and the transfer of electromagnetic energy into potential energy, the intensity of the CMBR, measured on earth, has been much lower than the real intensity has been at the origin.

As an example a visible universe like a sphere has been chosen with a radius of 4.4 10^{26} [m] and the earth located in the middle between the origin of the Big Bang and the outer boundaries of the visible Universe. As an example, two basic calculations have been done. The first one, calculating the total electromagnetic mass in the universe without taking into account the Electromagnetic-Gravitational Interaction.

$$M_{UNIVERSE} = \int_{R_1}^{R_2} \rho_{EM} \, 4 \, \pi \, r^2 \, d\,r = \int_{0.0113}^{4.4\times10^{26}} \frac{5\times10^{-36}}{\left(r \,/\, 2.2\times10^{26}\right)^2} 4 \, \pi \, r^2 d\,r = \qquad (6\text{-f})$$

$$= 1.338 \; 10^{45} \; [kg]$$

The second calculation has been done by taking into account the Electromagnetic-Gravitational Interaction, including the Electromagnetic-Gravitational Conversion term

$$C_{EMGC} = e^{\frac{G\,m_1\,\varepsilon_0\,\mu_0}{r}}$$

$$M_{UNIVERSE} = \int_{R_1}^{R_2} \rho_{EM} \, e^{\frac{1}{r}} \, 4 \, \pi r^2 \, d\,r = \int_{0.0113}^{4.4\times10^{26}} \frac{5\times10^{-36}}{\left(r /2.2\times10^{26}\right)^2} \frac{e^{\frac{1}{r}}}{e^{\frac{1}{2.2\times10^{26}}}} \, 4 \, \pi \, r^2 d\,r = \qquad (6\text{-g})$$

$$= 1.077\times10^{53} \; [kg]$$

Including the effect of Electromagnetic-Gravitational Conversion changes an almost negligible effect of the CMBR on the total mass of the Universe into an effect that can easily explain the total mass of the Universe while neglecting the influences of the mass of the galaxies. And makes the Theory of Dark Matter avoidable.

3. Electromagnetic Radiation within a Spherical Coordinate System

The Spherical Coordinate System $\{r,\theta,\varphi,t\}$ is parameterized by the radius r of the Sphere, the polar angle θ and the azimuthal angle φ and the time t.

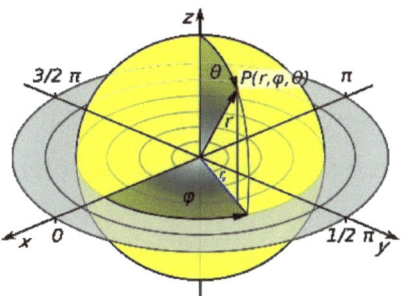

Figure 3. Spherical Coordinate System

The required <u>Electromagnetic Field Configuration</u> for a perfect Equilibrium in Space and Time follows from equation (5) and equals in Spherical Coordinates $\{r,\theta,\varphi,t\}$ for the Electric Field Components $e(\theta,r,\varphi,t)$:

$$\begin{pmatrix} e_r \\ e_\theta \\ e_\varphi \end{pmatrix} = \begin{pmatrix} 0 \\ \dfrac{1}{r}f(\theta,\varphi)g\left(t-\left(\dfrac{K_1}{r}+1\right)r\sqrt{\varepsilon_0\ \mu_0}\right) \\ 0 \end{pmatrix} \qquad (8)$$

The required Electromagnetic Field Configuration for a perfect Equilibrium in Space and Time follows from equation (4) and equals in Spherical Coordinates $\{r,\theta,\varphi,t\}$ for the Magnetic Field Components $m(\theta,r,\varphi,t)$:

$$\begin{pmatrix} m_r \\ m_\theta \\ m_\varphi \end{pmatrix} = \sqrt{\frac{\varepsilon_0}{\mu_o}} \begin{pmatrix} 0 \\ 0 \\ \dfrac{1}{r} f(\theta,\varphi) g\left(t - \left(\dfrac{K_1}{r} + 1 \right) r \sqrt{\varepsilon_0\ \mu_0} \right) \end{pmatrix}$$

For the divergence-free function $f(\theta,\varphi) = 1$, the solutions (8) and (9) are also the solutions for the known Maxwell Equations. For the non-divergence-free functions $f(\theta,\varphi)$, the solutions (8) and (9) are no solutions for the Maxwell Equations, which require divergence-free electromagnetic waves in the absence of any matter. [29,35,36,37,38]. They are however solutions of the <u>DEE (5)</u> and clearly they do exist in physics. Like the radiation of an inhomogeneous point light source like a LED.

4. Confined Electromagnetic Radiation within a Spherical Coordinate System through Electromagnetic-Gravitational Interaction

In physics it has been in generally assumed that the speed of light $c = 1/\sqrt{\varepsilon_0 \mu_0}$ is a physical constant. In this paragraph the possibilities will be discussed of a variable speed of light [10,11,12]. that can vary from zero until values higher than c. The only requirement for the existence of an Electromagnetic Field Configuration will be the requirement of a perfect equilibrium in space-time for the chosen electromagnetic field configuration[13,14,15]. This single unique requirement will always be a solution of the DEE (5).

The required <u>Electromagnetic Field Configuration</u> for a perfect Equilibrium in Space and Time[16,17,18,30,31,39] in respectively the : θ-direction $(f_\theta = 0)$ and the $\varphi - direction : (f\varphi = 0)$ follows from equation (5).

In Spherical Coordinates $\{r, \theta, \varphi, t\}$ the solution for the <u>DEE</u> <u>(5)</u> for the Electric Field Components $e(\theta, r, \varphi, t)$ equals:

$$\begin{pmatrix} e_r \\ e_\theta \\ e_\varphi \end{pmatrix} = \begin{pmatrix} 0 \\ f(r)g(\theta)h(\varphi)\operatorname{Sin}(\omega t) \\ -f(r)g(\theta)h(\varphi)\operatorname{Cos}(\omega t) \end{pmatrix} \qquad (10)$$

In Spherical Coordinates $\{r, \theta, \varphi, t\}$ the solution for the DEE (5) for the Magnetic Field Components $m(\theta, r, \varphi, t)$ in respectively the : θ-direction $(f_\theta = 0)$ and the

50

$\varphi - direction : \left(f\varphi = 0 \right)$ for the magnetic field components follows from equation (5) [29,30,31]. and equals:

$$
\begin{pmatrix} m_r \\ m_\theta \\ m_\varphi \end{pmatrix} = \begin{pmatrix} 0 \\ f(r)g(\theta)h(\varphi)\mathrm{Cos}(\omega t) \\ f(r)g(\theta)h(\varphi)\mathrm{Sin}(\omega t) \end{pmatrix} \qquad (11)
$$

Equation (4) gives the 3-dimensional force density f^a of an Electro-Magnetic Field Configuration[19,20,21]. in a coordinate free vector equation.

$$
\begin{pmatrix} f_r \\ f_\theta \\ f_\varphi \end{pmatrix} = \begin{pmatrix} \dfrac{2}{r}\,\varepsilon_0\,g(\theta)^2\,h(\varphi)^2\,f(r)\left[f(r) + r\,\dfrac{d\,f(r)}{d\,r} \right] \\ 0 \\ 0 \end{pmatrix} \qquad (12)
$$

It follows from equation (4) that the radiation pressure in radial direction does not counterbalance and does not equal zero.

The energy-density w_m of the Electromagnetic Configuration is essential for the calculation of the inward bounded gravitational pressure. The electromagnetic field configuration (10) and (11) for the functions $g(\theta) = 1$ and $h(\varphi) = 1$, results into the electromagnetic energy-density w_{em} :

51

$$\begin{pmatrix} e_r \\ e_\theta \\ e_\varphi \end{pmatrix} = \begin{pmatrix} 0 \\ f(r)\ \text{Sin}(\omega\, t) \\ -f(r)\ \text{Cos}(\omega\, t) \end{pmatrix} \qquad \begin{pmatrix} m_r \\ m_\theta \\ m_\varphi \end{pmatrix} = \begin{pmatrix} 0 \\ f(r)\ \text{Cos}(\omega\, t) \\ f(r)\ \text{Sin}(\omega\, t) \end{pmatrix}$$

(13)

$$W_{em} = \left(\frac{\mu_0}{2}\left(\overline{m}\cdot\overline{m}\right) + \frac{\varepsilon_0}{2}\left(\overline{e}\cdot\overline{e}\right) \right) = \varepsilon_0\, f(r)^2$$

According to Einstein's mass-energy equivalent $W = m\, c^2$,the specific electromagnetic mass[22,23,24,29]. density ρ_{em} equals:

$$\rho_{em} = \frac{1}{c^2}\left(\frac{\mu_0}{2}\left(\overline{m}\cdot\overline{m}\right) + \frac{\varepsilon_0}{2}\left(\overline{e}\cdot\overline{e}\right) \right) = \varepsilon_0{}^2\,\mu_0\, f(r)^2 \qquad (14)$$

The total electromagnetic mass [25,6,27,29,38]. M_{em} within a sphere with radius R equals:

$$M_{em} = 4\,\pi\,\varepsilon_0{}^2\mu_0 \int_0^R r^2\, f(r)^2\, d\,r \qquad (15)$$

At a distance r from the center of the sphere, the total electromagnetic mass M_{em} within the sphere[25,26,27,30]. causes, according Newton's Shell Theorem[28,29,30] , a gravitational field strength g_{em}:

$$g_{em} = \frac{4}{r^2}\,\pi\,\varepsilon_0^2\mu_o\, G_1 \int_0^R r^2\, f(r)^2\, dr \qquad (16)$$

In which G_1 is the gravitational constant and equals $G_1 = 6.67408\ 10^{-11}\ [\text{m}^3\ \text{kg}^{-1}\ \text{s}^{-2}]$. The gravitational inwards bounded radiation pressure[29,31,32,33]. follows from (16):

$$f_{GRAV} = \rho_{em}\, g_{em} = \frac{4}{r^2}\pi\, \varepsilon_0^4\, \mu_0^2\, G_1\, f(r)^2 \int_0^R r^2\, f(r)^2\, dr \qquad (17)$$

When there is a perfect equilibrium between the outwards bounded electromagnetic radiation pressure f_{RAD} and the inward bounded gravitational pressure [29,34,35,38,39]. f_{GRAV} we find from (12) and (17) the radius of the boundary sphere of the enclosed radiation $R_{BOUNDARY}$. for the functions $g(\theta) = 1$ and $h(\varphi) = 1$,

$$f_{RAD} = \frac{2}{r}\, \varepsilon_0 f(r)\left(f(r) + r\, \frac{d\, f(r)}{d\, r}\right) = \qquad (18)$$

$$= \frac{4}{r^2}\, \pi\varepsilon_0^4\, \mu_0^2\, G_1\, f(r)^2 \int^{R-BOUMNDARY} r^2 f(r)^2\, dr = f_{GRAV}$$

In the following example we choose for the function f(r)

$$f(r) = K_1\, r^n \qquad (19)$$

53

Substituting (19) in (18) results in the equation for $R_{BOUNDARY}$:

$$\frac{8.019056 \times 10^{-94} \ K1^6 \ R_{BOUNDARY}^{1+6n}}{3+2n} = 1.7708376 \times 10^{-11} \ K1^2 \left(1+n\right) R_{BOUNDARY}^{-1+2n} \quad (20)$$

Table 1:

Values for n	Values for K_1	$R_{BOUNDARY}$ [m]	Frequency ω_0	Area
-10	1	0.00072422 7	2.6×10^{12}	
-4	1	$5.5940239 \times 10^{-10}$	3.37×10^{18}	
-2	1	$6.7798267 \times 10^{-28}$	2.78×10^{36}	Electromagnetic Particle (Electromagnetically Controlled)
-2	2	$1.3559653 \times 10^{-27}$	1.39×10^{36}	
-2	10^{18}	$6.7798267 \times 10^{-10}$	2.78×10^{18}	
-1.6	1	$3.0622921 \times 10^{-45}$	6.15×10^{53}	
-1.51	1	$4.7888625 \times 10^{-52}$	3.94×10^{60}	
-1.51	10^{18}	$9.4264902 \times 10^{-17}$	2.0×10^{25}	
-1	1	Infinite		Solution of (5)
-0.9	1	$1.2126075 \times 10^{267}$	1.55×10^{-258}	Electromagnetic Black Hole (Gravitationally Controlled)
-0.5	1	2.1755186×10^{54}	8.66×10^{-46}	
-0.5	10^{18}			
-0.5	10^{36}			

55

5.   The fundamental conflict between Causality and Probability

The beginning of the conflict between Causality and Probability in Physics started at the historic invitation-only "Conseil Solvay" in 1911. Since that conference Albert Einstein has always defended the fundamental concept of Causality and the logical grounds for Causality and Effect while Niels Bohr has always defended the fundamental concept of Probability in which there is no relationship between Causality and Effect. The most fundamental and famous conference was the October 1927 "Fifth Solvay International Conference on Electrons and Photons" where the newly formulated Quantum Theory, based on Probability, had been accepted. Since then Quantum Physics has grown in power and has always been grounded on the material waves, for the first time mathematically described by Erwin Schrödinger and designated to be probability waves.

In Ref. (29) page 206 the "law of continuity" for electromagnetic radiation has been presented in equation (42), By presenting the electromagnetic field in a complex configuration in equation (48) on page 207, the electromagnetic continuity equation (42) has been presented as the quantum mechanical Schrödinger wave equation (53) on page 207. However the function ψ in (53) does not represent the quantum mechanical probability function but represents the confined electromagnetic field with harmonic frequency ω_0 in which the magnetic part has been presented as the real function $\overline{B} / \mu$ and the electric part has been presented as the imaginary function $i\,\overline{E} / c$. With "i" the imaginary number $\sqrt{-1}$. However as well as the electric part as the magnetic part of the confined electromagnetic field are both real.

The quantum mechanical wave function ψ is in real a vector function $\overline{\phi}$ which equals:

$$\overline{\phi} = \frac{1}{\sqrt{2\,\mu}} \left(\overline{B} + i\, \frac{\overline{E}}{c} \right) \tag{21.a}$$

and the complex conjugated vector function equals:

$$\overline{\phi}{}^* = \frac{1}{\sqrt{2\,\mu}} \left(\overline{B} - i\, \frac{\overline{E}}{c} \right) \tag{21.b}$$

And the dot product equals the electromagnetic energy density w:

$$\overline{\phi} \cdot \overline{\phi}{}^* = \frac{1}{2\,\mu} \left(\overline{B} + i\, \frac{\overline{E}}{c} \right) \cdot \left(\overline{B} - i\, \frac{\overline{E}}{c} \right) = \frac{1}{2}\, \mu\, H^2 + \frac{1}{2}\, \varepsilon\, E^2 = w \tag{21.c}$$

The cross product is proportional to the Poynting vector (ref. 29, page 202, equation 15).

$$\overline{\phi} \times \overline{\phi}{}^* = \frac{1}{2\,\mu} \left(\overline{B} + i\, \frac{\overline{E}}{c} \right) \times \left(\overline{B} - i\, \frac{\overline{E}}{c} \right) = i\,\sqrt{\varepsilon\,\mu}\ \overline{E} \times \overline{H} = i\,\sqrt{\varepsilon\,\mu}\ \overline{S} \tag{21.d}$$

In Ref. (29) on page 208 the "law of continuity" for electromagnetic radiation has been presented in equation (57). The vector function $\overline{\phi}$ represents the confined electromagnetic field with harmonic frequency ω_0 in which the magnetic part has been presented as the real vector function $\overline{B}$ and the electric part has been presented as the imaginary vector function $i\, \overline{E}\, /\, c$. By substituting the complex vector function $\overline{\phi}$ in (57) and using the relativistic

57

Lorentz transformations, the quantum mechanical relativistic Dirac equation has been derived in equation (102) on page 213. This has been published in 1995 in Physics Essays in A Continuous model of Matter (*DOI: 10.13140/RG.2.2.25149.77281*). (ref. 29).

This fundamental conflict is still going on and in this article mathematical results are presented in Table 1 in which the famous De Broglie waves, the material waves, designated as probability waves are electromagnetic waves with one harmonic frequency ω, confined by electromagnetic-gravitational interaction. Electromagnetic waves fully grounded on Causality and Effect.

Table 1 presents several values for the calculated Equilibrium Radius $R_{BOUNDARY}$ (20) at different values K_1 and n in Equation (18), (19) and (20) for the harmonic Electromagnetic-Gravitationally confined Electromagnetic waves with frequency ω_0:

6. Confined Electromagnetic Radiation within a Toroidal Coordinate System

The Toroidal Coordinate System $\{\theta, r, \varphi, R, t\}$ is parameterized by the large radius R of the Torus. The Toroidal Coordinate System is obtained by rotating bipolar coordinates $\{r, \varphi\}$ around an axis perpendicular to the axis connecting the two foci. The coordinate $\{\theta\}$ specifies the angle of rotation.

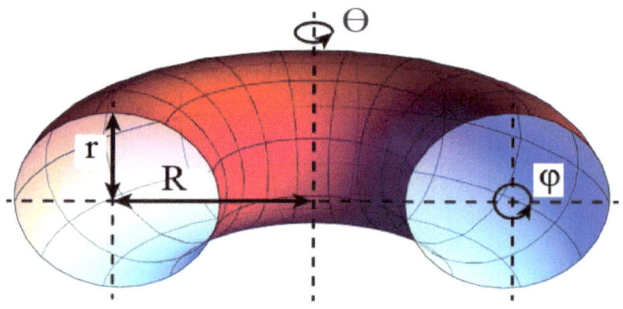

Figure 4. Toroidal Coordinate System

The required <u>Electromagnetic Field Configuration</u> for a perfect Equilibrium in Space and Time equals in Toroidal

Coordinates $\{\theta, r, \varphi, t\}$ for <u>the Electric Field Components e</u>
(θ, r, φ, t) .

$$
\begin{pmatrix} e_\theta \\ e_r \\ e_\varphi \end{pmatrix} = \left[\begin{array}{c} (0) \\[2em] \left(\dfrac{\dfrac{1}{\sqrt{2}} \operatorname{Csch}(r)(\operatorname{Cosh}(r) - \operatorname{Cos}(\theta))^{3/2} \ \sqrt{K1 - f_{13}(\theta, r, \varphi, t)^2}}{\sqrt{h(\theta) - 2 Cot(\theta) \operatorname{Tan}^{-1}\left(\operatorname{Tan}\left(\dfrac{\theta}{2}\right) \operatorname{Coth}\left(\dfrac{r}{2}\right) \right) + r}} \right) \\[3em] \left(\dfrac{\operatorname{Csch}(r)(\cosh(r) - \cos(\theta)) f_{13}(\theta, r, \varphi, t)}{\sqrt{-K2 + \dfrac{1}{2}\operatorname{Cos}(\theta)\left[h(\theta) - 2Cot(\theta)\operatorname{Tan}^{-1}\left(\operatorname{Tan}\left(\dfrac{\theta}{2}\right)\operatorname{Coth}\left(\dfrac{r}{2}\right)\right) + r \right] + h(\theta)}} \right. \\ \left. - \dfrac{1}{2}\operatorname{Cosh}(r)\left[h(\theta) - 2Cot(\theta)\operatorname{Tan}^{-1}\left(\tan\left(\dfrac{\theta}{2}\right)\coth\left(\dfrac{r}{2}\right)\right) + r \right] + \operatorname{Sinh}(r) \right) \end{array} \right]
$$

(22.a)

The required Electromagnetic Field Configuration for a perfect Equilibrium in Space and Time equals in Toroidal Coordinates $\{\theta,r,\varphi,t\}$ for <u>the Magnetic Field Components m</u> <u>(θ,r,φ,t)</u>.

$$
\begin{pmatrix} m_\theta \\ m_r \\ m_\varphi \end{pmatrix} = \sqrt{\frac{\varepsilon_0}{\mu_0}}
\begin{bmatrix}
(0) \\[2em]
\left(\dfrac{\dfrac{-1}{\sqrt{2}} \operatorname{Csch}(r)(\operatorname{Cosh}(r) - \operatorname{Cos}(\theta))^{3/2} \; f_{13}(\theta,r,\varphi,t)}{\sqrt{h(\theta) - 2Cot(\theta)\operatorname{Tan}^{-1}\left(\operatorname{Tan}\left(\dfrac{\theta}{2}\right)\operatorname{Coth}\left(\dfrac{r}{2}\right)\right) + r}} \right) \\[3em]
\left(\dfrac{\operatorname{Csch}(r)\,(\cosh(r) - \cos(\theta))\,\sqrt{K1 - f_{13}(\theta,r,\varphi,t)^2}}{\sqrt{-K2 + \dfrac{1}{2}\operatorname{Cos}(\theta)\left(h(\theta) - 2Cot(\theta)\operatorname{Tan}^{-1}\left(\operatorname{Tan}\left(\dfrac{\theta}{2}\right)\operatorname{Coth}\left(\dfrac{r}{2}\right)\right) + r\right) + h(\theta)}} \\ - \dfrac{1}{2}\operatorname{Cosh}(r)\left(h(\theta) - 2Cot(\theta)\operatorname{Tan}^{-1}\left(\tan\left(\dfrac{\theta}{2}\right)\coth\left(\dfrac{r}{2}\right)\right) + r\right) + \operatorname{Sinh}(r) \right)
\end{bmatrix}
\qquad (22.b)
$$

7. Confined Electromagnetic Radiation within a Toroidal Coordinate System through Electromagnetic-Gravitational Interaction in a non-linear Space-Time Continuum

The required <u>Electromagnetic Field Configuration</u> for a Gravitational-Electromagnetic Equilibrium in Space and Time equals in Toroidal Coordinates $\{\theta, r, \varphi, t\}$ for the Electric Field Components $e(\theta, r, \varphi, t)$:

$$
\begin{pmatrix} e_\theta \\ e_r \\ e_\varphi \end{pmatrix} = \begin{bmatrix} (0) \\ \left(\mathrm{Csch}(r)\ f1(\theta,r)\ (\mathrm{Cosh}(r) - \mathrm{Cos}(\theta))^{3/2}\ \sqrt{K1\text{-}f13(\theta,r,\varphi,t)^2}\ \right) \\ \dfrac{\mathrm{Csch}(r)\ (\mathrm{Cosh}(r) - \mathrm{Cos}(\theta))\ f13(\theta,r,\varphi,t)}{\sqrt{\mathrm{Cos}(\theta)\ f1(\theta,r)^2 - \mathrm{Cosh}(r)\ f1(\theta,r)^2 - g(r)}} \end{bmatrix} \quad (23)
$$

The required Electromagnetic Field Configuration for a Gravitational-Electromagnetic Equilibrium in Space and Time equals in Toroidal Coordinates $\{\theta, r, \varphi, t\}$ for the Magnetic Field Components $m(\theta, r, \varphi, t)$:

$$
\begin{pmatrix} m_\theta \\ m_r \\ m_\varphi \end{pmatrix} = \sqrt{\frac{\varepsilon_0}{\mu_0}} \begin{bmatrix} (0) \\ \left(\mathrm{Csch}(r)\ f1(\theta,r)\ \left(-(\mathrm{Cosh}(r) - \mathrm{Cos}(\theta))^{3/2} \right)\ f13(\theta,r,\varphi,t)\ \right) \\ \left(\dfrac{\mathrm{Csch}(r)\ (\mathrm{Cosh}(r) - \mathrm{Cos}(\theta))}{\sqrt{\mathrm{Cos}(\theta)\ f1(\theta,r)^2 - \mathrm{Cosh}(r)\ f1(\theta,r)^2 - g(r)}\ \sqrt{K1 - f13(\theta,r,\varphi,t)^2}} \right) \end{bmatrix} \quad (24)
$$

The toroidal electromagnetic field configuration is in perfect equilibrium with itself and its surrounding in respectively the θ- and the φ- direction. There is a resulting electromagnetic outward bounding force density in the r-direction, $\overrightarrow{J}(\upsilon,r,\psi,t)$ indicated as the outward bounding radiation pressure of the toroidal electromagnetic confinement.

$$\overrightarrow{J}(\upsilon,r,\varphi,t) = \begin{bmatrix} 0 \\ -\dfrac{\varepsilon_0 K_1 \text{Csch}^2(r) (\cos(\theta) - \cosh(r))^3 \left(4f1(\theta,r) f1^{(0,1)}(\theta,r) (\cos(\theta) - \cosh(r)) - g'(r) \right)}{2R} \\ 0 \end{bmatrix} \qquad (25)$$

This resulting outward bounding radiation pressure has to be compensated by the inward bounding gravitational force density, to create the required equilibrium by electromagnetic-gravitational interaction.

In a comparable way as in the example presented in spherical coordinates in (13) and (14) , the electromagnetic mass-density from the energy density in the torus can be calculated. With these values the inward bounded gravitational radiation pressure can be derived. From the requirement that both force densities have to compensate each other, the Boundary Radius $R_{BOUNDARY}$ for the toroidal confinement can be calculated.

8. The Origin of Electromagnetic Mass (Inertia)

When Erwin Schrödinger published in 1926 the well-known Schrödinger wave equation with the characteristic spherical and elliptical wave solutions, he found a mathematical presentation for "De Broglie-" or the Material Waves. When Bohr assigned these " De Broglie" waves as "Probability Waves", he excluded "De Broglie-Waves" from any mass. The mass was for 100 % in the elementary particle itself and "De Broglie Waves" without any mass described the probability of the position of this elementary particle. A complete non logical approach.

A fundamental logical approach is to assign the mass to "De Broglie Waves". And that is only possible when we "De Broglie Waves" are just "Confined Single Harmonic Electromagnetic Waves" (confined light). For this reason the mass of confined electromagnetic radiation has to be calculated. It will be clear that it is impossible to assign mass to free electromagnetic radiation. Because it is impossible to accelerate free electromagnetic radiation and the concept of mass if based on the inertia of mass by Newton's second law. For this reason the inertia of "Confined Single Harmonic Electromagnetic Waves" (confined light) has to be calculated.

We measure the mass in [kg] by acceleration of the object according Newton's second law of motion. In the theory of special relativity, the speed of light is a fundamental constant. But the intensity of the light will not be constant. Speed is relative. When we emit by a laser a beam of light, the intensity of the light will not change with the distance (without any divergence of the beam).

However there is a difference. When the speed of the observer has the same speed as the speed of the light source, then the observer and the light source are relative at rest. And the same light intensity will be measured at the location of the emitter and at the location of the observer.

When the observer moves towards the emitter, the intensity of the light at the location of the observer will increase with $\gamma \, (1 + v / c)$ according the Lorentz transformation in which "v" is the relative velocity between emitter and observer . At low velocities the term γ will equal 1.

When the observer moves away from the emitter, the intensity of the light at the location of the observer will decrease with $\gamma \, (1 - v / c)$ according the Lorentz transformation. At low velocities the term γ will equal 1.

When light is confined between two 100 % reflecting mirrors, then we can conclude that the speed of both mirrors will always be equal, relative to each other. And at uniform speed, the radiation pressures on both mirrors will be equal and both opposite directed radiation pressures will neutralize.

During acceleration, it will take time for the light to travel with the speed of light between both mirrors. When we consider one mirror as the emitter and the opposite mirror as the observer, we can conclude that the speed of the emitter will be different (when the beam of light leaves the emitter) than the speed of the observer (when the beam of light reaches the observer) because of the time interval, needed for the beam of light to propagate from emitter to observer during the acceleration.

During the acceleration, both opposite oriented radiation pressures on both mirrors will not be equal anymore and they will not neutralize each other anymore. During acceleration, there will be a resulting force according Newton's second law of motion caused by both different radiation pressures.

For the first step in this calculation an imaginary experiment has been used. Two 100 % reflecting mirrors B and A (both in the x-y plane and without any mass) are placed opposite each other at a distance Δ x (ref 29, page 7, figure 1). A single harmonic electromagnetic wave has been confined between both mirrors. Between both mirrors a "Standing Electromagnetic Wave" appears which has been formed by two waves travelling in opposite directions along the z-axis.

The Poynting vector corresponding with the electromagnetic wave propagating along the z-axis in the + direction (positive direction of the z-axis) has been indicated as $\overline{S}^+ = \overline{E}^+ \times \overline{H}^+$ and the Poynting vector corresponding with the electromagnetic wave propagating along the z-axis in the - direction (opposite direction) has been indicated as $\overline{S}^- = \overline{E}^- \times \overline{H}^-$.

The system is at rest. The radiation pressures, caused by the confined electromagnetic radiation, on both mirrors A and B are opposite and equal in magnitude:

$$P_A = \frac{2\,S_A}{c} = \frac{2\,S_B}{c} = P_B \qquad (26)$$

Einstein has formulated this very well. "Velocities are always relative" . To calculate the radiation pressure on Mirror A, the velocities, only relative to Mirror A for the waves with the respective Poynting vectors $\overline{S}^{+} = \overline{E}^{+} \times \overline{H}^{+}$ and $\overline{S}^{-} = \overline{E}^{-} \times \overline{H}^{-}$, have to be calculated.

8.1.1 The radiation pressure on Mirror A, when Mirror A moves with a velocity v in the direction of the positive z-axis

When the system of "Two Mirrors B - A" moves in the direction of the positive z-axis, Mirror A moves in the direction of the positive z-axis and the Poynting vector $\overline{S}^+ = \overline{E}^+ \times \overline{H}^+$ will decrease according the Lorentz transformation (ref. 29, page 23, A-57).

$$\overline{S}_v^+ = \overline{E}_v^+ \times \overline{H}_v^+ = \gamma^2 \left(1 - \frac{v}{c}\right)^2 \left(\overline{E}^+ \times \overline{H}^+\right) \qquad (27)$$

When the system of "Two Mirrors B - A" moves in the direction of the positive z-axis, Mirror A moves in the direction of the positive z-axis the Poynting vector $\overline{S}^- = \overline{E}^- \times \overline{H}^-$ will increase according the Lorentz transformation (ref. 29, page 222, A-57).

$$\overline{S}_v^- = \overline{E}_v^- \times \overline{H}_v^- = \gamma^2 \left(1 + \frac{v}{c}\right)^2 \left(\overline{E}^+ \times \overline{H}^+\right) \qquad (28)$$

The total radiation pressure, caused by the confined electromagnetic radiation, on mirror A equals:

$$P_A = \frac{S_A^+ + S_A^-}{c} = \frac{\gamma^2 \left[\left(1 - \frac{v}{c}\right)^2 + \left(1 + \frac{v}{c}\right)^2\right]\left(\overline{E}^+ \times \overline{H}^+\right)}{c} \qquad (29)$$

68

8.1.2 The radiation pressure on Mirror B when Mirror B moves with a velocity v in the direction of the positive z-axis

When the system of "Two Mirrors B - A" moves in the direction of the positive z-axis, Mirror B moves in the direction of the positive z-axis and the Poynting vector $\overline{S}^- = \overline{E}^- \times \overline{H}^-$ will increase according the Lorentz transformation (ref. 29, page 23, A-57).

$$\overline{S}_v^- = \overline{E}_v^- \times \overline{H}_v^- = \gamma^2 \left(1 + \frac{v}{c} \right)^2 \left(\overline{E}^+ \times \overline{H}^+ \right) \quad (3$$

When the system of "Two Mirrors B - A" moves in the direction of the positive z-axis, Mirror A moves in the direction of the positive z-axis the Poynting vector $\overline{S}^+ = \overline{E}^+ \times \overline{H}^+$ will increase according the Lorentz transformation (ref. 29, page 23, A-57).

$$\overline{S}_v^+ = \overline{E}_v^+ \times \overline{H}_v^+ = \gamma^2 \left(1 - \frac{v}{c} \right)^2 \left(\overline{E}^+ \times \overline{H}^+ \right) \quad (3$$

The total <u>radiation pressure</u>, caused by the confined electromagnetic radiation, on mirror B equals:

$$P_A = \frac{S_A^+ + S_A^-}{c} = \frac{\gamma^2 \left(\left(1 + \frac{v}{c} \right)^2 + \left(1 - \frac{v}{c} \right)^2 \right) \left(\overline{E}^+ \times \overline{H}^+ \right)}{c} \quad ($$

P_A and P_B are still equal in magnitude and both in opposite direction and still cancel each other. The system fulfils Newton's first law of motion.

8.2 Newton's second Law of Motion (Inertia) for Confined Electromagnetic Radiation

When the system of "Two Mirrors B - A" accelerates, the velocity increases with Δv in a time interval Δt. At time t the radiation pressures on mirror A and mirror B are presented in (29) and (32). At time $t + \Delta t$ the radiation pressures on Mirror A and Mirror B will different:

The <u>radiation pressure</u> at time $t + \Delta t$ caused by the confined electromagnetic radiation, on mirror A equals:

$$P_A = \frac{S_A^+ + S_A^-}{c} = \frac{\gamma^2 \left[\left(1 + \frac{(v)}{c}\right)^2 + \left(1 - \frac{(v + \Delta v)}{c}\right)^2 \right] \left(\overline{E}^+ \times \overline{H}^+\right)}{c} \quad (3$$

Because the wave with Poynting vector $\overline{S}^+ = \overline{E}^+ \times \overline{H}^+$ has left Mirror B at "t" and during the time interval Δt the magnitude of $\overline{E}_t^+ = \left(1 + \frac{v}{c}\right)\overline{E}^+$ and $\overline{H}_t^+ = \left(1 + \frac{v}{c}\right)\overline{H}^+$ has not changed.

The <u>radiation pressure</u> at time $t + \Delta t$ caused by the confined electromagnetic radiation, on mirror B equals:

$$P_B = \frac{S_B^+ + S_B^-}{c} = \frac{\gamma^2 \left[\left(1 + \frac{(v + \Delta v)}{c}\right)^2 + \left(1 - \frac{(v)}{c}\right)^2 \right] \left(\overline{E}^+ \times \overline{H}^+\right)}{c} \quad (34)$$

71

Because the wave with Poynting vector $\overline{S} = \overline{E} \times \overline{H}$ has left Mirror A at "t" and during the time interval Δt the magnitude of $\overline{E_t} = \left(1 + \dfrac{v}{c}\right)\overline{E}$ and $\overline{H_t} = \left(1 + \dfrac{v}{c}\right)\overline{H}$ has not changed.

The radiation pressures on Mirror A and Mirror B do not counterbalance each other anymore and the resulting radiation pressure equals:

$$P_B - P_A = \frac{\gamma^2 (4\,\Delta v)\,S}{c^2} \tag{35}$$

Equation (35) can be written as:

$$P_B - P_A = \frac{\gamma^2 (4\,\Delta v)\,S}{c^2} = \frac{\gamma^2\left(4\,\dfrac{\Delta v}{\Delta t}\right) S \Delta t}{c^2} = \gamma^2\,\frac{W}{c^2}\,a = \gamma^2\,m\,a \tag{36}$$

In which the acceleration $a = \dfrac{\Delta v}{\Delta t}$ and the inertia $m = \dfrac{W}{c^2}$.

At non relativistic velocities $\gamma = 1$ and (36) turns into the Newton's second law of motion. From (36) also Einstein's famous equation $W = m\,c^2$ follows. In (36) W is the total confined electromagnetic mass.

Now we can consider electromagnetic confinements without mirrors but electromagnetic confinements through electromagnetic gravitational interaction. By superposition and integration over arbitrary surfaces it is possible to prove that all confined electromagnetic radiation equals (36) and has electromagnetic mass and follow Newton's second law of motion.

9. Quantum Mechanical Entanglement

One of the first answers every Quantum Physicist will give you when you doubt the Holy Grail of the "Particle-Wave Duality" in Quantum Physics is Bell's Theorem. But the "Particle-Wave Duality" can never explain the laws in Quantum Mechanics. In special the "Particle-Wave Duality" cannot explain "Entanglement".

In the new model of the "Particle-Wave-Mass" Tri-Unity , the concept of a particle on itself has completely disappeared. A particle itself does not exist anymore. The "Particle-Wave-Mass" is a wave of confined light (Electromagnetic Radiation) with a "particle aspect" and a "mass aspect". And sometimes we observe only one of the three aspects.

To demonstrate this, a simple model of a string in a music instrument like a piano, will be used. As soon the string has been excited, we will hear a tune (frequency spectrum). Along the string a pattern of "standing (stationary) transversal waves" occurs. When we observe one single harmonic frequency we observe that along the string a pattern occurs of nodes and anti-nodes occurs. The nodes as well as the anti-nodes are connected together in an "Entanglement" And because the information travels along the string simultaneously in both directions, it gives the impression that the information travels with an infinite speed. Because the sound wave in the string has been confined between the two ends of the string, Entanglement occurs.

Entanglement is only possible within the confinement of wave patterns. Because only then the wave pattern can mathematically be considered as the superposition of waves propagating in opposite directions and forming "standing (stationary) waves" It does not matter if the "standing (stationary) waves" occur because of the confinement by the ends of the string or a confinement of an electromagnetic field due to "Electromagnetic-Gravitational Interaction".

Now we consider the creation of an "electron-anti electron" pair. Both have been created at the same location. That is the exact location where the string has been excited in the example. When the electron-anti electron pair moves away from each other, it is still in the essence "one confined wave pattern". And only through Gravitational-Electromagnetic Interaction this wave pattern has been confined and the area of confinement becomes only larger when both particles move apart. The electron/anti-electron pair is still one stationary confined harmonic electromagnetic wave pattern, which is the superposition of a complex nodes-anti nodes pattern. Confined by its own "Electromagnetic-Gravitational Field "Entanglement" will occur in a comparable way how it occurs in the string of a music instrument. Confinement is the necessary requirement for the existence of "Entanglement".

Entanglement is the fundamental evidence that the "Particle-Wave Duality" in Classical Quantum Mechanics is not able to explain "Entanglement". Particles can not be at the same place simultaneously. The weakness in Classical Quantum Mechanics is the "particle". Not the "wave". Modern experiments crumble down the "Particle-Wave-Duality" model of Niels Bohr more and more. How is it possible to make a photo of a complex probability wave? How is "Entanglement" possible? How can a photon create an electron/anti-electron pair?

Classical Quantum Physics needs a make-over. A new concept that does not crumble down further by every new experiment. And the particle concept is the weakness in Classical Quantum Mechanics. We come closer and closer to the final conclusion that particles do not exist. "Elementary particles" is a concept that worked very well for the last 2000 years. But it does no work anymore. We have to give up a way of thinking in the way we are already thinking for more than 2000 years since Plato introduced the concept of "Elementary particle".

Times are changing and it is time for us to leave the ancient save roads of the Greek Philosophers and give up the safety of our illusions. It is time to move on and let the "particle" go.

Let us build a new model. A new model without particles. A model free from the ancient Greek Philosophers. A model just grounded only on the wave.

And that is what I am offering. A new model without particles. In which the property of matter will be carried only by the wave. And the wave is already given to us since the creation of our Universe. The Light Wave. The Light Wave which carries the ability to confine itself and manifest itself like a particle. The Light wave which carries the ability to confine itself and manifest itself in the property of inertia and carries mass. The confined light wave that is in fundamental essence just only a wave but carries the 3-unity in itself of particle -wave – mass. The Particle-Wave-Mass 3-Unity.

10.1 A Gravitational-Electromagnetic model beyond the Superstring

Calculations in **Mathematica 11.0 .nb** file and in **PDF file**.

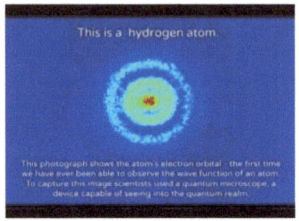

Figure 5 First Image of the Hydrogen Atom's Orbital structure

"De Broglie Waves" are real and do exist. Schrödinger" and as well Dirac both have written the simple well-known (electromagnetic) Continuity Equation (ref. 29, page 208, equation 57). in a complex form. What they have really found are electromagnetic waves. Single harmonic and confined. The Schrödinger Solution of a Spherical Probability wave around the nucleus has been interpreted wrong. It is not a complex probability wave. It is a real electromagnetic wave in which the real part is the solution for the electric part and the imaginary part is the magnetic part, just written with an "i" index before the term. The quantum mechanical wave function ψ is in real a vector function $\overline{\phi}$ which equals:

$$\overline{\phi} = \frac{1}{\sqrt{2\,\mu}} \left(\overline{B} + i\, \frac{\overline{E}}{c} \right) \tag{3}$$

and the complex conjugated vector function equals:

$$\overline{\phi}* = \frac{1}{\sqrt{2\,\mu}} \left(\overline{B} - i\, \frac{\overline{E}}{c} \right) \tag{3}$$

And the dot product equals the electromagnetic energy density w:

$$\bar{\phi} \cdot \bar{\phi}^* = \frac{1}{2\mu} \left(\bar{B} + i \frac{\bar{E}}{c} \right) \cdot \left(\bar{B} - i \frac{\bar{E}}{c} \right) = \frac{1}{2} \mu H^2 + \frac{1}{2} \varepsilon E^2 = w \qquad (39\text{-a})$$

The cross product is proportional to the Poynting vector (ref. 29, page 202, equation 15).

$$\bar{\phi} \times \bar{\phi}^* = \frac{1}{2\mu} \left(\bar{B} + i \frac{\bar{E}}{c} \right) \times \left(\bar{B} - i \frac{\bar{E}}{c} \right) = i \sqrt{\varepsilon \mu} \, \bar{E} \times \bar{H} = i \sqrt{\varepsilon \mu} \, \bar{S} \qquad (39\text{-b})$$

The Gravitational-Electromagnetic Confinement for the elementary structure beyond the "superstring" is presented in equation (5-a).

$$- \frac{1}{c^2} \frac{\partial (\bar{E} \times \bar{H})}{\partial t} + \varepsilon_0 \, \bar{E} \, (\nabla \cdot \bar{E}) - \varepsilon_0 \, \bar{E} \times (\nabla \times \bar{E}) + \mu_0 \, \bar{H} \, (\nabla \cdot \bar{H}) - \qquad (5\text{-a})$$

$$- \mu_0 \, \bar{H} \times (\nabla \times \bar{H}) - \frac{1}{2} \varepsilon_0^{\,2} \mu_0 \left(\bar{E} \cdot \bar{E} \right) \bar{g} - \frac{1}{2} \varepsilon_0 \mu_0^{\,2} \left(\bar{H} \cdot \bar{H} \right) \bar{g} = \bar{0}$$

In which $\bar{g}$ represents the (radial oriënted) gravitational acceleration caused by the electromagnetic mass density of the confined electromagnetic radiation.

The solution for equation (5-a) equals:

$$\begin{pmatrix} e_r \\ e_\theta \\ e_\varphi \end{pmatrix} = \begin{pmatrix} 0 \\ f(r)\ \text{Sin}(\omega t) \\ -f(r)\ \text{Cos}(\omega t) \end{pmatrix} \qquad \begin{pmatrix} m_r \\ m_\theta \\ m_\varphi \end{pmatrix} = \begin{pmatrix} 0 \\ f(r)\ \text{Cos}(\omega t) \\ f(r)\ \text{Sin}(\omega t) \end{pmatrix}$$

(40)

$$W_{em} = \left(\frac{\mu_0}{2} \left(\overline{m} \cdot \overline{m} \right) + \frac{\varepsilon_0}{2} \left(\overline{e} \cdot \overline{e} \right) \right) = \varepsilon_0\, f(r)^2$$

In which f(r) equals:

$$f[r] = K\ e^{-\dfrac{-\dfrac{G1\ \varepsilon_0\ \mu_0}{r} + 8\,\pi\, \log[r]}{8\,\pi}}$$

(41)

10.1.1 A Gravitational-Electromagnetic Confinement Type 1
(emm = 10⁻⁴ [kg]; radius = 2 x 10⁻³⁵ [m]):

The chosen values equal:

$$f[r] = K\ e^{-\dfrac{-\dfrac{G1\ \text{emm}\ \varepsilon_0\ \mu_0}{r} + 8\ \pi\ \log[r]}{8\ \pi}}$$

$G1 = 6.6740810^{-11}$

emm $= 10^{-4}$

$\varepsilon_0 = 8.8510^{-12}$

$\mu_0 = 1.256637061435917210^{-6}$

(42)

In which "emm" equals the electromagnetic mass of the confinement located at the center according Newton's Shell Theorem.

For an electromagnetic mass of the confinement type (1): emm = 10⁻⁴ [kg], the radius of the confinement equals approximately 2 x 10⁻³⁵ [m]. This is in the order of Planck's Length,

The Plot graph of the Electric Field Intensity f(r) of the confinement has been presented as a function of the radius in figure (7) and figure (8):

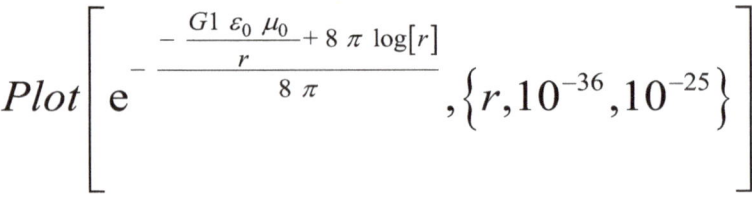

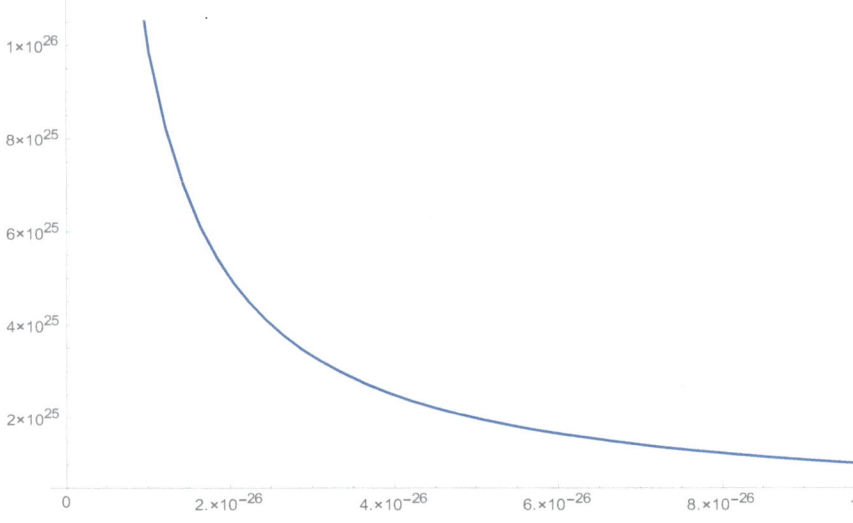

Figure 6   PlotGraph of the Electric Field Intensity f(r) for the region $10^{-36} < r < 10^{-25}$ in which the gravitational field acceleration has been chosen accordingly an electromagnetic mass of 10^{-4} [kg] located at the center of the confinement, according Newton's Shell Theorem.

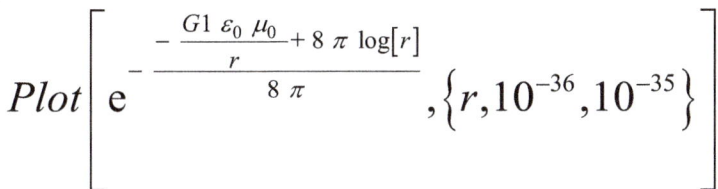

$$Plot\left[e^{-\dfrac{-\dfrac{G1\,\varepsilon_0\,\mu_0}{r}+8\,\pi\,\log[r]}{8\,\pi}}, \left\{r,10^{-36},10^{-35}\right\}\right]$$

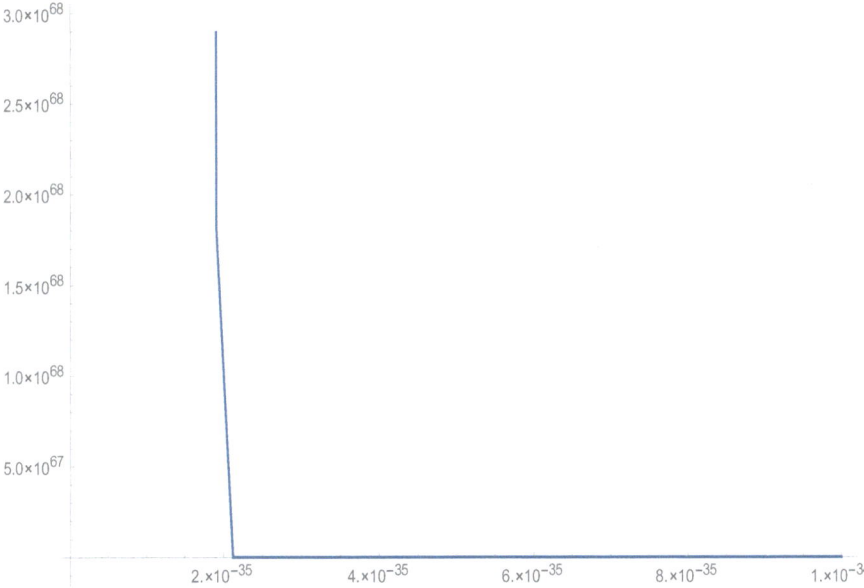

Figure 7   PlotGraph of the Electric Field Intensity f(r) for the region $10^{-36} < r < 10^{-35}$ in which the gravitational field acceleration has been chosen accordingly an electromagnetic mass of 10^{-4} [kg] located at the center of the confinement, according Newton's Shell Theorem.

It follows from Figure 8 that the radius of the stable gravitational electromagnetic confinement equals approximately 2 x 10^{-35} [m], which is the size of the Planck length. According the theory of superstrings, the fundamental constituents of reality are strings of the Planck length (about 1.62 10^{-35} [m]) that vibrate at resonant frequencies.

10.1.2 A Gravitational-Electromagnetic Confinement Type 2 (emm = 10^{-12} [kg]; radius = 2.5 x 10^{-43} [m]):

For an electromagnetic mass of the confinement type (2): emm = 10^{-12} [kg], the radius of the confinement equals approximately 2.5 x 10^{-43} [m]. This is much smaller than Planck's Length and has been indicated as "sub Planck Length".

The Plot graph of the Electric Field Intensity f(r) of the confinement has been presented as a function of the radius in figure (9)_ and figure (10):

$$Plot\left[e^{-\dfrac{-\dfrac{G1\,\varepsilon_0\,\mu_0}{r}+8\,\pi\,\log[r]}{8\,\pi}} , \{r,10^{-43},10^{-40}\} \right]$$

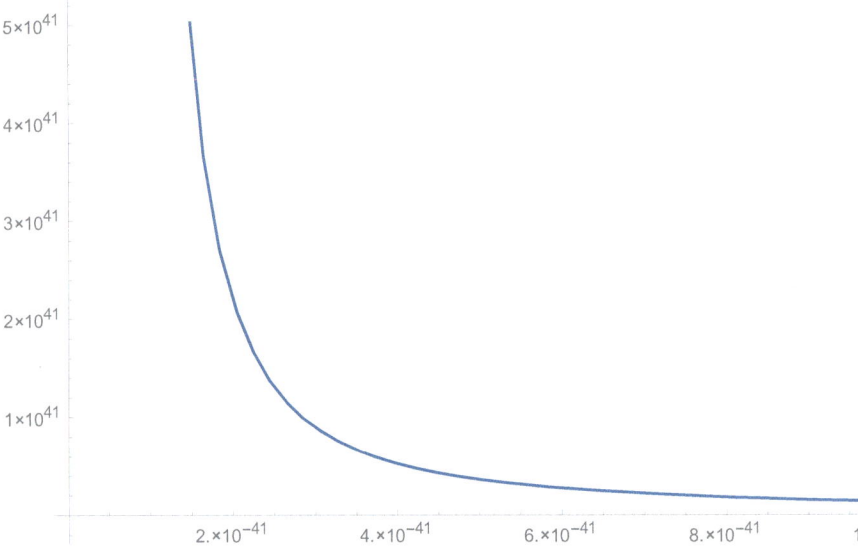

Figure 8   PlotGraph of the Electric Field Intensity f(r) for the region $10^{-36} < r < 10^{-25}$ in which the gravitational field acceleration has been chosen accordingly an electromagnetic mass of 10^{-4} [kg] located at the center of the confinement, according Newton's Shell Theorem.

85

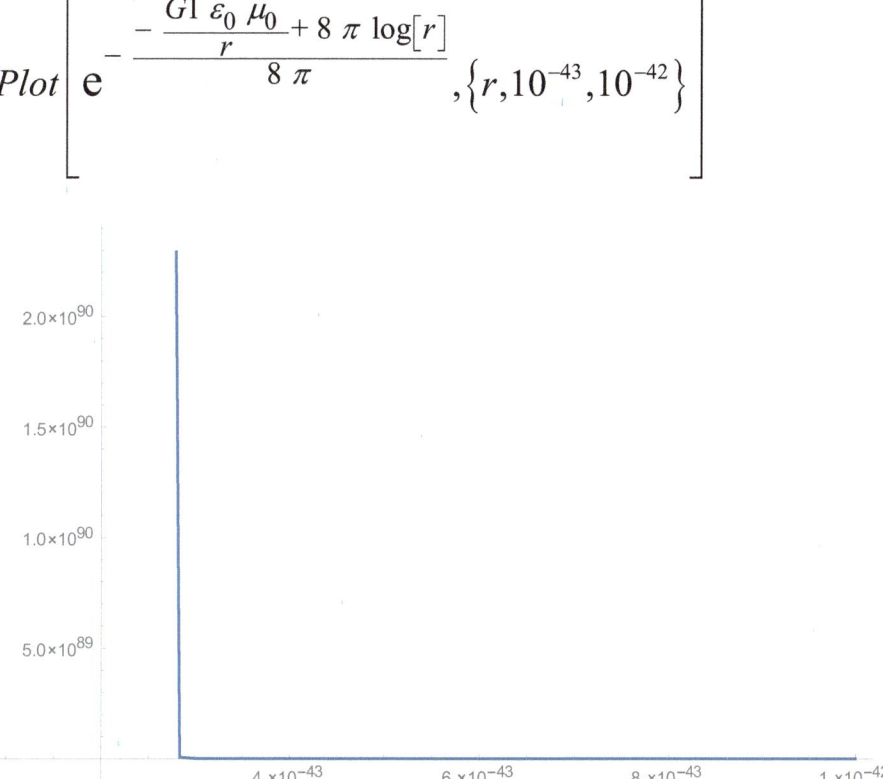

$$Plot\left[e^{-\dfrac{-\dfrac{G1\,\varepsilon_0\,\mu_0}{r}+8\,\pi\,\log[r]}{8\,\pi}}, \left\{ r, 10^{-43}, 10^{-42} \right\} \right]$$

Figure 9   PlotGraph of the Electric Field Intensity f(r) for the region $10^{-43} < r < 10^{-42}$ in which the gravitational field acceleration has been chosen accordingly an electromagnetic mass of 10^{-12} [kg] located at the center of the confinement, according Newton's Shell Theorem.

86

10.1.3 A Gravitational-Electromagnetic Confinement
Type 3 (emm = 1.6726 x 10^{-27} [kg]; radius = 3 x 10^{-58} [m]):

For an electromagnetic mass of the confinement type (3):
emm = 1.6726 x10^{-27} [kg] (mass of proton), the radius of the
confinement equals approximately 3 x 10^{-58} [m]. This is far
beyond the order of Planck's Length,

The Plot graph of the Electric Field Intensity f(r) of the
confinement has been presented as a function of the radius in
figure (11) and figure (12):

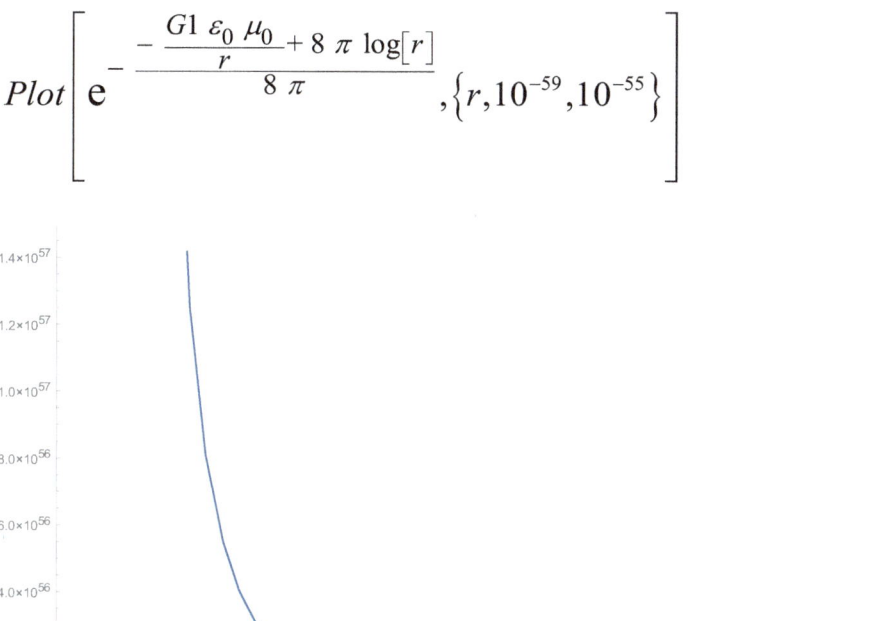

$$Plot\left[e^{-\dfrac{-\dfrac{G1\,\varepsilon_0\,\mu_0}{r}+8\,\pi\,\log[r]}{8\,\pi}},\left\{r,10^{-59},10^{-55}\right\}\right]$$

Figure 10   PlotGraph of the Electric Field Intensity f(r) for the region $10^{-59} < $ r $< 10^{-55}$ in which the gravitational field acceleration has been chosen accordingly an electromagnetic mass of 1.6 726 x 10^{-27} [kg] located at the center of the confinement, according Newton's Shell Theorem.

$$Plot\left[e^{-\dfrac{-\dfrac{G1\,\varepsilon_0\,\mu_0}{r}+8\,\pi\,\log[r]}{8\,\pi}},\left\{r,10^{-59},10^{-57}\right\}\right]$$

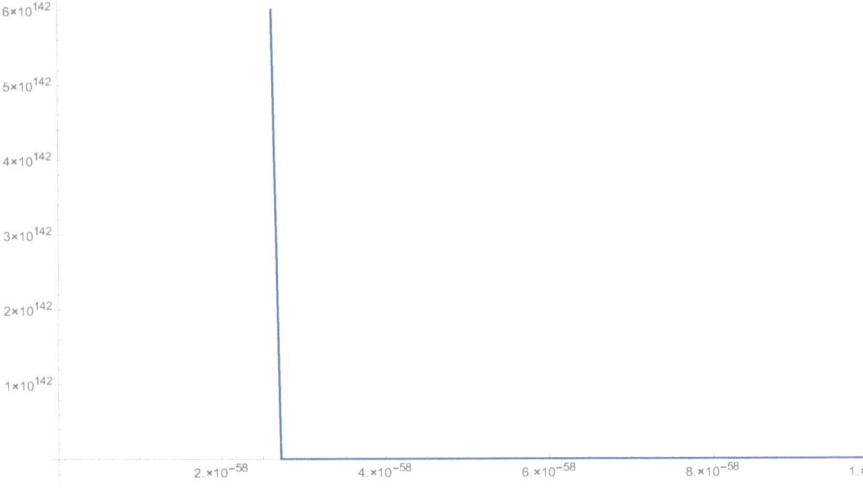

Figure 11   PlotGraph of the Electric Field Intensity f(r) for the region $10^{-59} < r < 10^{-57}$ in which the gravitational field acceleration has been chosen accordingly an electromagnetic mass of 1.6726×10^{-27} [kg] located at the center of the confinement, according Newton's Shell Theorem.

The fundamental question is: How it is possible to create confinements from "visible light" (with a wave length between 3.9×10^{-7} [m] until 7×10^{-7} [m]) within dimensions smaller than Planck's Length?

This is only possible when the wave length of the confined radiation is smaller than de dimensions of the confinement. This requires extreme high frequencies. The transformation in frequency from visible light into the extreme high frequency of the confinement is possible because of the Lorentz transformation during the collapse of the radiation when the confinement has been formed (implosion of visible light).

89

10.2 The Origin of Electric Charge and Magnetic Spin in discrete values

The Gravitational-Electromagnetic Confinement for the elementary structure of the "superstring" is presented in equation (5-a).

$$- \frac{1}{c^2} \frac{\partial \, (\overline{E} \times \overline{H})}{\partial t} + \varepsilon_0 \, \overline{E} \, (\nabla \, . \, \overline{E}) - \varepsilon_0 \, \overline{E} \times (\nabla \times \overline{E}) + \mu_0 \, \overline{H} \, (\nabla \, . \, \overline{H}) - \quad (5\text{-a})$$

$$- \mu_0 \, \overline{H} \times (\nabla \times \overline{H}) - \frac{1}{2} \, \varepsilon_0^2 \mu_0 \left(\overline{E} \, . \, \overline{E} \right) \, \overline{g} - \frac{1}{2} \, \varepsilon_0 \mu_0^2 \left(\overline{H} . \, \overline{H} \right) \, \overline{g} = \overline{0}$$

In which $\overline{g}$ represents the (radial oriented) gravitational acceleration caused by the electromagnetic mass density of the confined electromagnetic radiation.

To find the origin of Electric Charge and Magnetic Spin we choose as an example the solution for equation (5-a) which equals:

$$\begin{pmatrix} e_r \\ e_\theta \\ e_\varphi \end{pmatrix} = \begin{pmatrix} 0 \\ f1(r,\theta,\varphi,t) \; \mathrm{Sin}(\omega \, t) \\ - f2(r,\theta,\varphi,t) \; \mathrm{Cos}(\omega \, t) \end{pmatrix} \qquad \begin{pmatrix} m_r \\ m_\theta \\ m_\varphi \end{pmatrix} = \begin{pmatrix} 0 \\ f2(r,\theta,\varphi,t) \; \mathrm{Cos}(\omega \, t) \\ f1(r,\theta,\varphi,t) \; \mathrm{Sin}(\omega \, t) \end{pmatrix}$$

$$\tag{43}$$

$$w_{em} = \left(\frac{\mu_0}{2} \left(\overline{m} \, . \, \overline{m} \right) + \frac{\varepsilon_0}{2} \left(\overline{e} \, . \, \overline{e} \right) \right) = \varepsilon_0 \, f(r)^2$$

In which $f[r]$, $f1[r,\theta,\varphi,t]$, $f2[r,\theta,\varphi,t]$ equals:

$$f[r] = K\, e^{-\dfrac{-\dfrac{G1\,\varepsilon_0\,\mu_0}{r} + 8\,\pi\,\log[r]}{8\,\pi}}$$

$$f1[r,\theta,\varphi,t] = K\, e^{-\dfrac{-\dfrac{G1\,\varepsilon_0\,\mu_0}{r} + 8\,\pi\,\log[r]}{8\,\pi}}\, g1[\theta,\varphi,t] \qquad (4$$

$$f2[r,\theta,\varphi,t] = \dfrac{K\, e^{-\dfrac{-\dfrac{G1\,\varepsilon_0\,\mu_0}{r} + 8\,\pi\,\log[r]}{8\,\pi}}\,\sqrt{-g1[\theta,\varphi,t]^2 + \cos[2\,t\,\omega]\,g1[\theta,\varphi,t]^2 + 2h[\theta,\varphi]}}{\sqrt{2}}$$

In which $g1[\theta,\varphi,t]$ and $h[\theta,\varphi]$ are arbitrary function.

The "sub Max Planck's length" confinement has been described for the electric field intensity:

$$\begin{pmatrix} e_r \\ e_\theta \\ e_\varphi \end{pmatrix} = \begin{pmatrix} 0 \\[2ex] K\, e^{-\dfrac{-\dfrac{G1\,\varepsilon_0\,\mu_0}{r} + 8\,\pi\,\log[r]}{8\,\pi}}\, g1[\theta,\varphi,t]\sin[t\omega] \\[2ex] -\dfrac{K\, e^{-\dfrac{-\dfrac{G1\,\varepsilon_0\,\mu_0}{r} + 8\,\pi\,\log[r]}{8\,\pi}}\,\sqrt{-g1[\theta,\varphi,t]^2 + \cos[2t\omega]\,g1[\theta,\varphi,t]^2 + 2h[\theta,\varphi]}}{\sqrt{2}} \end{pmatrix} \qquad (45)$$

The "sub Max Planck's length" confinement has been described for the magnetic field intensity:

$$\begin{pmatrix} m_r \\ m_\theta \\ m_\varphi \end{pmatrix} = \sqrt{\dfrac{\varepsilon_0}{\mu_0}}\begin{pmatrix} 0 \\[2ex] \dfrac{K\, e^{-\dfrac{-\dfrac{G1\,\varepsilon_0\,\mu_0}{r} + 8\,\pi\,\log[r]}{8\,\pi}}\,\sqrt{-g1[\theta,\varphi,t]^2 + \cos[2t\omega]\,g1[\theta,\varphi,t]^2 + 2h[\theta,\varphi]}}{\sqrt{2}} \\[2ex] K\, e^{-\dfrac{-\dfrac{G1\,\varepsilon_0\,\mu_0}{r} + 8\,\pi\,\log[r]}{8\,\pi}}\, g1[\theta,\varphi,t]\sin[t\omega] \end{pmatrix} \qquad (46)$$

91

10.3 The 5 Types of "Sub Max Planck Length Gravitational-Electromagnetic Confinements" resulting in Electric Charge and Magnetic Spin

The following functions with the quantum variables {m1, n1, p1, q1} have been chosen:

$$f[r] = K\, e^{-\dfrac{-\dfrac{G1\,\varepsilon_0\,\mu_0}{r} + 8\,\pi\,\log[r]}{8\,\pi}}$$

$$g1(\theta,\varphi,t) \;=\; \sin(t\,\omega)\,(\sin(\pi\,\theta\,\text{m1})\sin(\text{n1}\,2\,\pi\,\varphi) + 1)$$

$$h(\theta,\varphi) \;=\; \sin(\pi\,\theta\,\text{p1})\,\sin(\text{q1}\,2\,\pi\,\varphi) + 1 \tag{47}$$

$$g2(\theta,\varphi,t) \;=\; \frac{\sec(t\,\omega)\,\sqrt{\cos(2\,t\,\omega)\,g1(\theta,\varphi,t)^2 \;-\; g1(\theta,\varphi,t)^2 \;+\; 2\,h(\theta,\varphi)}}{\sqrt{2}}$$

$$f1[r,\theta,\varphi,t] \;=\; e^{-\dfrac{-\dfrac{G1\,\varepsilon_0\,\mu_0}{r} + 8\,\pi\,\log[r]}{8\,\pi}}\,K\,g1[\theta,\varphi,t]$$

$$f2[r,\theta,\varphi,t] \;=\; \frac{e^{-\dfrac{-\dfrac{G1\,\varepsilon_0\,\mu_0}{r} + 8\,\pi\,\log[r]}{8\,\pi}}\,K\,\sqrt{-g1[\theta,\varphi,t]^2 + \cos[2\,t\,\omega]\,g1[\theta,\varphi,t]^2 + 2h[\theta,\varphi]}}{\sqrt{2}}$$

10.3.1 Type 1 of "Sub Max Planck Length Gravitational-Electromagnetic Confinements" (Electric- and Magnetic Dipoles, Electric- and Magnetic Spin) {m1=0, n1=0, p1=0, q1=0}

The divergence of the electric field intensity (electric charge density) equals:

$$\nabla \cdot \begin{pmatrix} e_r \\ e_\theta \\ e_\varphi \end{pmatrix} = \frac{\sqrt{2}\; K1\; \cot(\theta)\; \sin^2(t\;\omega)\; \sqrt{1 - \sin^4(t\;\omega)}\; e^{\frac{G_1\,\varepsilon_0\,\mu_0}{8\,\pi\,r}}}{r^2 \sqrt{2 - 2\sin^4(t\omega)}} \tag{48}$$

$$\nabla \cdot \begin{pmatrix} e_r \\ e_\theta \\ e_\varphi \end{pmatrix} = \frac{\frac{1}{2}\; K1\; \cot(\theta)\; e^{\frac{G_1\,\varepsilon_0\,\mu_0}{8\,\pi\,r}}}{r^2} \quad \text{(averaged over 1 period of time)}$$

The divergence of the magnetic field intensity (magnetic monopole) equals:

$$\nabla \cdot \begin{pmatrix} m_r \\ m_\theta \\ m_\varphi \end{pmatrix} = \frac{K1\; \sqrt{\varepsilon_0}\; \cot(\theta)\; \sqrt{2 - 2\sin^4(t\;\omega)}\; e^{\frac{G_1\,\varepsilon_0\,\mu_0}{8\,\pi\,r}}}{\sqrt{2}\; \sqrt{\mu 0}\; r^2} \tag{4*}$$

$$\nabla \cdot \begin{pmatrix} m_r \\ m_\theta \\ m_\varphi \end{pmatrix} = \frac{K1\; \sqrt{\varepsilon_0}\; \cot(\theta)\; \sqrt{\frac{3}{4}}\; e^{\frac{G_1\,\varepsilon_0\,\mu_0}{8\,\pi\,r}}}{\sqrt{\mu 0}\; r^2} \quad \text{(averaged over 1 period of time)}$$

In which K1 is an arbitrary variable. Because of the $\cot(\theta)$ function, the electric divergence as well as the magnetic divergence changes from sign when the angle θ varies

93

between 0^0 until 360^0 forming electric dipoles (+ versus -) and magnetic dipoles (N versus S).

10.3.2 Type 2 of "Sub Max Planck Length Gravitational-Electromagnetic Confinements" (Electric- and Magnetic Dipoles, Electric- and Magnetic Spin) {m1=1, n1=0, p1=0, q1=0}

The divergence of the electric field intensity (electric charge density) equals:

$$\nabla \cdot \begin{pmatrix} e_r \\ e_\theta \\ e_\varphi \end{pmatrix} = \frac{\sqrt{2}\ K1\ \cot(\theta)\ \sin^2(t\omega)\ \sqrt{1 - \sin^4(t\omega)}\ e^{\frac{G_1\ \varepsilon_0\ \mu_0}{8\ \pi\ r}}}{r^2 \sqrt{2 - 2\ \sin^4(t\omega)}}$$

(50

$$\nabla \cdot \begin{pmatrix} e_r \\ e_\theta \\ e_\varphi \end{pmatrix} = \frac{\frac{1}{2} K1\ \cot(\theta)\ e^{\frac{G_1\ \varepsilon_0\ \mu_0}{8\ \pi\ r}}}{r^2} \qquad \text{(averaged over 1 period of time)}$$

The divergence of the magnetic field intensity (magnetic monopole) equals:

$$\nabla \cdot \begin{pmatrix} m_r \\ m_\theta \\ m_\varphi \end{pmatrix} = \frac{K1\ \sqrt{\varepsilon_0}\ \cot(\theta)\ \sqrt{2 - 2\ \sin^4(t\omega)}\ e^{\frac{G_1\ \varepsilon_0\ \mu_0}{8\ \pi\ r}}}{\sqrt{2}\ \sqrt{\mu_0}\ r^2}$$

(51)

$$\nabla \cdot \begin{pmatrix} m_r \\ m_\theta \\ m_\varphi \end{pmatrix} = \frac{K1\ \sqrt{\frac{3}{4}}\ \sqrt{\varepsilon_0}\ \cot(\theta)\ e^{\frac{G_1\ \varepsilon_0\ \mu_0}{8\ \pi\ r}}}{\sqrt{\mu_0}\ r^2} \qquad \text{(averaged over 1 period of time)}$$

In which K1 is an arbitrary variable. Because of the $\cot(\theta)$ function, the electric divergence as well as the magnetic divergence changes from sign when the angle θ varies between 0^0 until 360^0 forming electric dipoles (+ versus -) and magnetic dipoles (N versus S).

10.3.3 Type 3 of "Sub Max Planck Length Gravitational-Electromagnetic Confinements" {m1=1, n1=1, p1=0, q1=0}

The divergence of the electric field intensity (electric charge density) equals:

$$
\nabla \cdot \begin{pmatrix} e_r \\ e_\theta \\ e_\varphi \end{pmatrix} = \frac{K1 \ \sin^2(t\omega) \ e^{\frac{G_1 \, \varepsilon_0 \, \mu_0}{8 \, \pi \, r}} \left(\cos(\varphi)(\sin(\theta)\sin(\varphi)+1)\sin^2(t\omega) \right)}{r^2 \sqrt{1-(\sin(\theta)\sin(\varphi)+1)^2 \sin^4(t\omega)}} \ +
$$

$$
\frac{K1 \ \sin^2(t\omega) \ e^{\frac{G_1 \, \varepsilon_0 \, \mu_0}{8 \, \pi \, r}} \left((2\cos(\theta)\sin(\varphi)+\cot(\theta))\sqrt{1-(\sin(\theta)\sin(\varphi)+1)^2 \sin^4(t\omega)} \right)}{r^2 \sqrt{1-(\sin(\theta)\sin(\varphi)+1)^2 \sin^4(t\omega)}}
$$

$$(52)$$

$$
\nabla \cdot \begin{pmatrix} e_r \\ e_\theta \\ e_\varphi \end{pmatrix} = \frac{K1 \ e^{\frac{G_1 \, \varepsilon_0 \, \mu_0}{8 \, \pi \, r}} \left(\frac{1}{2}\cos(\varphi)(\sin(\theta)\sin(\varphi)+1) \right)}{2 \, r^2 \sqrt{1-\frac{1}{4}(\sin(\theta)\sin(\varphi)+1)^2}} \ +
$$

$$
\frac{K1 \ e^{\frac{G_1 \, \varepsilon_0 \, \mu_0}{8 \, \pi \, r}} \left((2\cos(\theta)\sin(\varphi)+\cot(\theta)) \ \sqrt{1-\frac{1}{4}(\sin(\theta)\sin(\varphi)+1)^2} \right)}{2 \, r^2 \sqrt{1-\frac{1}{4}(\sin(\theta)\sin(\varphi)+1)^2}} \quad \text{(averaged over 1 period of time)}
$$

The divergence of the magnetic field intensity (magnetic monopole) equals:

$$\nabla \cdot \begin{pmatrix} m_r \\ m_\theta \\ m_\varphi \end{pmatrix} = \frac{- K1 \sqrt{\varepsilon_0}\; e^{\frac{G_1 \varepsilon_0 \mu_0}{8\pi r}} \left(\sin(\varphi)\sin^4(t\omega)(\sin(2\theta)\sin(\varphi)) \right)}{\sqrt{\mu_0}\; r^2 \sqrt{1 - (\sin(\theta)\sin(\varphi) + 1)^2 \sin^4(t\omega)}} +$$

$$\frac{K1 \sqrt{\varepsilon_0}\; e^{\frac{G_1 \varepsilon_0 \mu_0}{8\pi r}} \left(3\cos(\theta)) + \cos(\varphi)\sin^2(t\omega)\sqrt{1 - (\sin(\theta)\sin(\varphi)+1)^2 \sin^4(t\omega)} - \cot(\theta)\left(\sin^4(t\omega) - 1\right) \right)}{\sqrt{\mu_0}\; r^2 \sqrt{1 - (\sin(\theta)\sin(\varphi)+1)^2 \sin^4(t\omega)}}$$

$$(53)$$

$$\nabla \cdot \begin{pmatrix} m_r \\ m_\theta \\ m_\varphi \end{pmatrix} = \frac{- K1 \sqrt{\varepsilon_0}\; e^{\frac{G_1 \varepsilon_0 \mu_0}{8\pi r}} \left(\sin(\varphi)\frac{1}{4}(\sin(2\theta)\sin(\varphi) + 3\cos(\theta)) \right)}{\sqrt{\mu_0}\; r^2 \sqrt{1 - \frac{1}{4}(\sin(\theta)\sin(\varphi) + 1)^2}} +$$

$$\frac{K1 \sqrt{\varepsilon_0}\; e^{\frac{G_1 \varepsilon_0 \mu_0}{8\pi r}} \left(\frac{1}{2}\cos(\varphi)\sqrt{1 - \frac{1}{4}(\sin(\theta)\sin(\varphi) + 1)^2} + \frac{3}{4}\cot(\theta) \right)}{\sqrt{\mu_0}\; r^2 \sqrt{1 - \frac{1}{4}(\sin(\theta)\sin(\varphi) + 1)^2}} \quad \text{(averaged over 1 period of time)}$$

In which K1 is an arbitrary variable with a Positive (positive charge) or a Negative (negative charge) value.

97

10.3.4 Type 4 of "Sub Max Planck Length Gravitational-Electromagnetic Confinements" (Electric- and Magnetic Dipoles, Electric- and Magnetic Spin) {m1=0, n1=0, p1=1, q1=0}

The divergence of the electric field intensity (electric charge density) equals:

$$\nabla \cdot \begin{pmatrix} e_r \\ e_\theta \\ e_\varphi \end{pmatrix} = \frac{e^{\frac{G_1 \, \varepsilon_0 \, \mu_0}{8 \, \pi \, r}} K1 \cot[\theta] \sin[t\omega]^2}{r^2}$$

(54)

$$\nabla \cdot \begin{pmatrix} e_r \\ e_\theta \\ e_\varphi \end{pmatrix} = \frac{e^{\frac{G_1 \, \varepsilon_0 \, \mu_0}{8 \, \pi \, r}} K1 \cot[\theta]}{2 \, r^2} \quad \text{(averaged over 1 period of time)}$$

In which K1 is an arbitrary variable with a Positive (positive charge) or a Negative (negative charge) value.

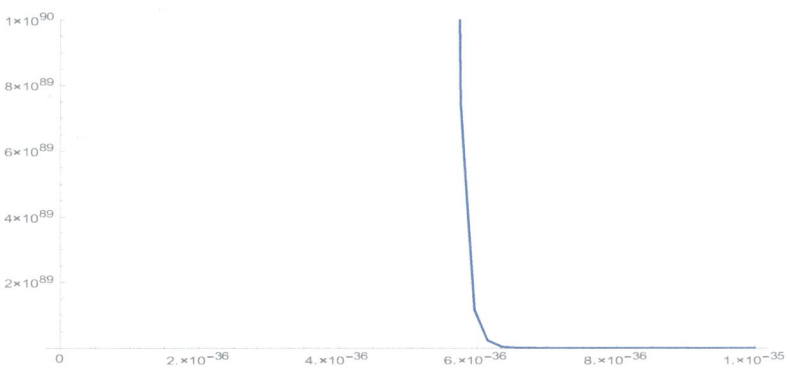

Figure 12. Equation (54). Averaged over 1 period of time. Electric Charge Density Plot in a range from 10^{-37} [m] until 10^{-35} [m].

The divergence of the magnetic field intensity (magnetic monopole) equals:

$$\nabla \cdot \begin{pmatrix} m_r \\ m_\theta \\ m_\varphi \end{pmatrix} = \frac{e^{\frac{G_1\, \varepsilon_0\, \mu_0}{8\,\pi\,r}}\; K1\, \sqrt{\varepsilon_0}\; \cot[\theta]\; \sqrt{1-\sin[t\omega]^4}}{r^2\, \sqrt{\mu_0}}$$

(55)

$$\nabla \cdot \begin{pmatrix} m_r \\ m_\theta \\ m_\varphi \end{pmatrix} = \frac{e^{\frac{G_1\, \varepsilon_0\, \mu_0}{8\,\pi\,r}}\; K1\, \sqrt{\varepsilon_0}\; \cot[\theta]}{2\, r^2\, \sqrt{\mu_0}} \quad \text{(averaged over 1 period of time)}$$

In which K1 is an arbitrary variable. Because of the $\cot(\theta)$ function, the electric divergence as well as the magnetic divergence changes from sign when the angle θ varies between 0^0 until 360^0 forming electric dipoles (+ versus -) and magnetic dipoles (N versus S).

10.3.5 Type 5 of "Sub Max Planck Length Gravitational-Electromagnetic Confinements" {m1=0, n1=0, p1=1, q1=1}

The divergence of the electric field intensity (electric charge density) equals:

$$\nabla \cdot \begin{pmatrix} e_r \\ e_\theta \\ e_\varphi \end{pmatrix} = \frac{e^{\frac{G_1\, \varepsilon_0\, \mu_0}{8\,\pi\,r}}\, K1 \left(2\,\cot[\theta]\,\sin[t\omega]^2 \;-\; \dfrac{\cos[\varphi]}{\sqrt{1 + \sin[\theta]\,\sin[\varphi] - \sin[t\omega]^4}} \right)}{2\,r^2}$$

(56)

$$\nabla \cdot \begin{pmatrix} e_r \\ e_\theta \\ e_\varphi \end{pmatrix} = \frac{e^{\frac{G_1\, \varepsilon_0\, \mu_0}{8\,\pi\,r}}\, K1 \left(\cot[\theta] \;-\; \dfrac{\cos[\varphi]}{\sqrt{\dfrac{3}{4} + \sin[\theta]\,\sin[\varphi]}} \right)}{2\,r^2} \qquad \text{(averaged over 1 period of time)}$$

The divergence of the magnetic field intensity (magnetic monopole) equals:

$$\nabla \cdot \begin{pmatrix} m_r \\ m_\theta \\ m_\varphi \end{pmatrix} = \frac{e^{\frac{G_1\, \varepsilon_0\, \mu_0}{8\,\pi\,r}}\, K1\, \sqrt{\varepsilon_0}\, \left(3\,\cos[\theta]\,\sin[\varphi] - 2\,\cot[\theta]\left(-1 + \sin[t\omega]^4\right) \right)}{2\,r^2\, \sqrt{\mu_0}\, \sqrt{1 + \sin[\theta]\,\sin[\varphi] - \sin[t\omega]^4}}$$

(57)

$$\nabla \cdot \begin{pmatrix} m_r \\ m_\theta \\ m_\varphi \end{pmatrix} = \frac{e^{\frac{G_1\, \varepsilon_0\, \mu_0}{8\,\pi\,r}}\, K1\, \sqrt{\varepsilon_0}\, \left(3\,\cos[\theta]\,\sin[\varphi] + \dfrac{3}{2}\,\cot[\theta] \right)}{2\,r^2\, \sqrt{\mu_0}\, \sqrt{\dfrac{3}{4} + \sin[\theta]\,\sin[\varphi]}} \qquad \text{(averaged over 1 period of time)}$$

In which K1 is an arbitrary variable with a Positive (positive charge) or a Negative (negative charge) value.

100

10.4 Type II of "Sub Max Planck Length Gravitational-Electromagnetic Confinements"

The "sub Max Planck's length" Type II confinement has been described for the electric field intensity:

$$
\begin{pmatrix} e_r \\ e_\theta \\ e_\varphi \end{pmatrix} = \begin{pmatrix} 0 \\ \dfrac{e^{\frac{G1\,\varepsilon_0\,\mu_0}{8\,\pi\,r}}\, h[\theta,\varphi]\, \sin[\omega\,t]^2 \sin\left[r\,\sqrt{\varepsilon_0\,\mu_0}\,\,\omega\right]^2}{r} \\ -\dfrac{e^{\frac{G1\,\varepsilon_0\,\mu_0}{8\,\pi\,r}}\, h[\theta,\varphi]\, \sqrt{K1 - \sin[\omega\,t]^4 \sin\left[r\,\sqrt{\varepsilon_0\,\mu_0}\,\,\omega\right]^4}}{r} \end{pmatrix}
\tag{58}
$$

The "sub Max Planck's length" confinement has been described for the magnetic field intensity:

$$
\begin{pmatrix} m_r \\ m_\theta \\ m_\varphi \end{pmatrix} = \sqrt{\frac{\varepsilon_0}{\mu_0}} \begin{pmatrix} 0 \\ \dfrac{e^{\frac{G1\,\varepsilon_0\,\mu_0}{8\,\pi\,r}}\, h[\theta,\varphi]\, \sqrt{K1 - \sin[\omega\,t]^4 \sin\left[r\,\sqrt{\varepsilon_0\,\mu_0}\,\,\omega\right]^4}}{r} \\ \dfrac{e^{\frac{G1\,\varepsilon_0\,\mu_0}{8\,\pi\,r}}\, h[\theta,\varphi]\, \sin[\omega\,t]^2 \sin\left[r\,\sqrt{\varepsilon_0\,\mu_0}\,\,\omega\right]^2}{r} \end{pmatrix}
\tag{59}
$$

The divergence of the electric field intensity (electric charge density) equals:

$$\nabla \cdot \begin{pmatrix} e_r \\ e_\theta \\ e_\varphi \end{pmatrix} = \frac{e^{\frac{G1e}{8\pi r}} \sqrt{K1 - \sin[t\ \omega]^4 \sin\left[r\sqrt{\varepsilon_0\ \mu_0}\ \omega\right]^4}\ h^{(0,1)}[\theta,\varphi]}{r}$$

$$+ \frac{e^{\frac{G1e}{8\pi r}}\ \sin[t\ \omega]^2 \sin\left[r\ \sqrt{\varepsilon_0\ \mu_0}\ \omega\right]^2 h^{(1,0)}[\theta,\varphi]}{r}$$

(60)

In which K1 is a positive constant equal or larger than 1.

The divergence of the magnetic field intensity (magnetic monopole) equals:

$$\nabla \cdot \begin{pmatrix} m_r \\ m_\theta \\ m_\varphi \end{pmatrix} = \frac{e^{\frac{G_1\ \varepsilon_0\ \mu_0}{8\ \pi\ r}} \sqrt{\varepsilon_0} \sin[t\ \omega]^2 \sin\left[r\ \sqrt{\varepsilon_0\ \mu_0}\ \omega\right]^2 h^{(0,1)}[\theta,\varphi]}{r\sqrt{\mu_0}}$$

$$+ \frac{e^{\frac{G_1\ \varepsilon_0\ \mu_0}{8\ \pi\ r}} \sqrt{\varepsilon_0} \sqrt{K1 - \sin[t\ \omega]^4 \sin\left[r\ \sqrt{\varepsilon_0\ \mu_0}\ \omega\right]^4} h^{(1,0)}[\theta,\varphi]}{r\sqrt{\mu_0}}$$

(61)

The function has been chosen:

$$h[\theta,\varphi] = \sin[n\ \theta]\cos[m\ \varphi]$$

(62)

In which the integers n = 0, ½, 1½, 2, 2½, 3, 3½,…. And m =0, ½, 1½, 2, 2½, 3, 3½,….:

$$\rho = \varepsilon_0 \nabla \cdot \begin{pmatrix} e_r \\ e_\theta \\ e_\varphi \end{pmatrix} = \frac{n\ \varepsilon_0\ e^{\frac{G1e}{8\pi r}}\ \cos(n\ \theta)\ \cos(m\ \varphi)\ \sin(t\ \omega)^2 \sin\left(r\ \sqrt{\varepsilon_0\ \mu_0}\ \omega\right)^2}{r}$$

(63)

According to Gauss's law the electric charge density ρ equals for m = 0 and n = 1 for an electric monopole.

Equation (63) represents a di-pole function for the electric charge density.

For the corresponding magnetic di-pole flux density ϕ (spin) equals for n = 0 and m = + ½ (spin up) and m = - ½ (spin down):

$$\phi = \mu_0 \nabla. \begin{pmatrix} m_r \\ m_\theta \\ m_\varphi \end{pmatrix} = \frac{m \sqrt{\varepsilon_0 \mu_0} \; e^{\frac{G1\epsilon}{8\pi r}} \; \cos(n\,\theta) \cos(m\,\varphi) \sin\left(r \sqrt{\varepsilon_0 \mu_0} \; \omega\right)^2}{r} \qquad (64)$$

Equation (64) represents a di-pole function for the magnetic flux density (spin).

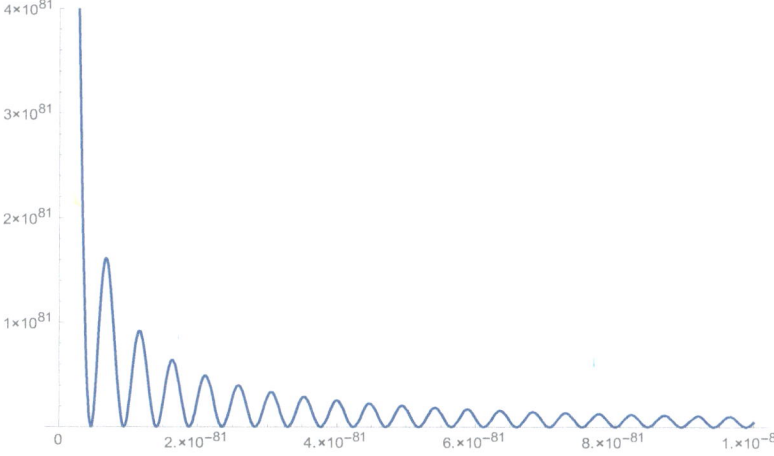

Figure 13 Equation (56): PlotGraph of the Electric Field Intensity f(r) for the region $10^{-85} < r < 10^{-80}$ with a frequency of $\omega = 10^{90}$ [s^{-1}] in which the gravitational field acceleration has been chosen accordingly an electromagnetic mass of 1.6726 x 10^{-27} [kg] located at the center of the confinement, according Newton's Shell Theorem.

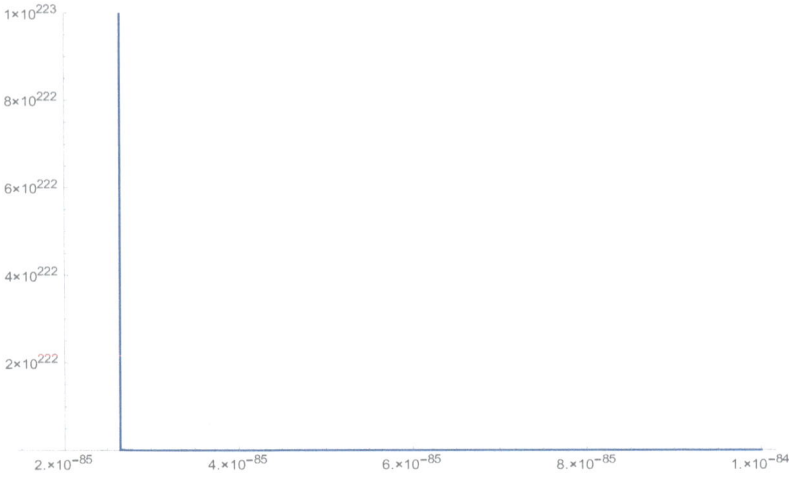

Figure 14 Equation (56): PlotGraph of the Electric Field
Intensity f(r) for the region $10^{-85} < r < 10^{-84}$ with a
frequency of $\omega = 10^{90}$ in which the gravitational field
acceleration has been chosen accordingly an electromagnetic
mass of 1.6726 x 10^{-27} [kg] located at the center of the
confinement, according Newton's Shell Theorem.

11. Concluding Remarks

The example of Gravitational-Electromagnetic Interaction, presented in table 1 shows two types of confinement.

1. For values $0 < n < -1$, the Gravitational-Electromagnetic Confinement will be Gravitationally controlled (Table 1). This means that for values for $r > R_{BOUNDARY}$ the inward bounded Gravitational for will be larger than the outward bounded Electromagnetic Radiation pressure. Electromagnetic Radiation will be attracted by Gravity towards the confinement at the surface $R_{BOUNDARY}$. Because for values $r < R_{BOUNDARY}$ the outward bounded radiation pressure is higher than the inward bounded gravitational pressure, all the radiation will be forced to be confined at equilibrium just at the surface of the spherical sphere with radius $R_{BOUNDARY}$. The confinement can be considered as an Electromagnetic Black Hole.

2. For values $-1 < n < -\infty$, the Gravitational-Electromagnetic Confinement will be Electromagnetically controlled (Table 1). This means that for values for $r > R_{BOUNDARY}$ the inward bounded Gravitational for will be smaller than the outward bounded Electromagnetic Radiation pressure. Electromagnetic Radiation will be scattered by the Radiation Pressure away from the confinement at the surface $R_{BOUNDARY}$. Because for values $r < R_{BOUNDARY}$ the outward bounded radiation pressure is smaller than the inward bounded gravitational pressure, all the radiation will be confined within the sphere with

radius $R_{BOUNDARY}$. The confinement can be considered as an Electromagnetic Particle.

3. For values $n = -1$, the inward bounded Gravitational pressure equals the outward bounded Electromagnetic Radiation pressure at any distance r. The calculated value for $R_{BOUNDARY}$ becomes $R_{BOUNDARY} \rightarrow \infty$.

Because of the extremely high-energy densities within electromagnetic-gravitational confinements and the extremely small dimensions, the radiation pressure at small densities will be extremely high. For this reason, single harmonic (monochromatic) electromagnetic-gravitational confinements will behave like nondeformable particles in experiments.

Table 1 demonstrates that the mass density division within the Electromagnetic Gravitational Confinement determines a wide range of diameters of the Confinements varying from 10^{-52} [m] (which is 10^{-42} times smaller than the diameter of the atom) until 10^{54} [m] (which is 10^{27} times larger than the diameter of the observable universe).

The exact solution for a Gravitational Electromagnetic Confinement results in diameters much smaller than Planck's Length ($1.616229 \ 10^{-35}$ [m]).

The fundamental question is: How it is possible to create confinements from "visible light" (with a wave length between 3.9×10^{-7} [m] until 7×10^{-7} [m]) within dimensions smaller than Planck's Length?

This is only possible when the wave length of the confined radiation is smaller than de dimensions of the confinement. This requires extreme high frequencies. The transformation in frequency from visible light into the extreme high frequency of the confinement is possible because of the Lorentz transformation during the collapse of the radiation when the confinement has been formed (implosion of visible light).

12. Three fundamental Equations in Qantum Physics

Taking the 4-dimensional Divergence of the 4-dimensional Stress Energy Tensor[1,16,29] (Equation 3) results in a 4-dimensional 4-vector with 4 components, which can be presented as 3 fundamentally different Electromagnetic/ Quantum Mechanical equations.

1) The 4th component of this 4-vector equals Poynting's Theorem. Evidence has been presented that the Schrödinger Wave Equation is a complex notation for Poynting's Theorem (equation 42, Ref. 29) rewritten in a complex way (equation 55, Ref.29)

2) The 4th component of this 4-vector equals Poynting's Theorem. Evidence has been presented that the relativistic Dirac Equation is a complex notation for Poynting's Theorem (equation 57, Ref. 29) rewritten in a complex way (equation 102, Ref.29)

3) The first 3 components of this 4-vector result in the 3-Dimensional Vector Equations (5) and (5-a) describing the "Unified 4-Dimensional Hyperspace Equilibrium" (beyond Einstein 4-Dimensional, Kaluza-Klein 5-Dimensional and Superstring 10- and 11 Dimensional Curved Hyperspaces). The fourth component of this 4-vector results in the relativistic Dirac Equation (5.P).

Acknowledgment

I specially would like to thank Professor Dr. F.W. Sluijter (1936-2014) TU/e, The Netherlands. He has shown a way in physics that has been forgotten by so many. A humble man man with a great wisdom. He has always been one of the happy few of the superior level of scientists like Albert Einstein who think completely independently. Albert Einstein, whose ideas have been the foundation of this work. A man of wisdom, independently thinking and a great insight in Physics. He had the courage and vision to create a new unknown path in physics to find beyond our understanding that what unifies the Universe.

I am indebted to all and to my family in particular, who always have supported me on a long and winding road in life and in physics.

References

1.   Li-Xin Li; A New Unified Theory of Electromagnetic and Gravitational Interactions, Frontiers of Physics, Volume 11, Issue 6, article id. 110402 (2016); arxiv.org/abs/1511.01260

2.   Richard Easther, Brian R Greene, Mark G Jackson and Daniel Kabat; String windings in the early universe, Journal of Cosmology and Astroparticle Physics, Volume 2005, February 2005

3.   J. Wheeler, Phys. Rev. 97, 511 (1955).

4.   Dirk Englund, Arka Majumdar, Michal Bajcsy, Andrei Faraon, Pierre Petroff, and Jelena Vučković; Ultrafast Photon-Photon Interaction in a Strongly Coupled Quantum Dot-Cavity System, Phys. Rev Lett. 108, 093604, March 2012, DOI : 10.1103/PhysRevLett.108.093604

5.   L. Filipe O. Costa, Georgios Lukes-Gerakopoulos, and Oldřich Semerák; Spinning particles in general relativity: Momentum-velocity relation for the Mathisson-Pirani spin condition; Phys. Rev. D 97, 084023 – Published 16 April 2018

6.   Ryotaro Kase, Masato Minamitsuji, and Shinji Tsujikawa; Relativistic stars in vector-tensor theories ; Phys. Rev. D 97, 084009 – Published 9 April 2018

7.   Hector O. Silva, Jeremy Sakstein, Leonardo Gualtieri, Thomas P. Sotiriou, and Emanuele Berti; Spontaneous Scalarization of Black Holes and Compact Stars from a Gauss-Bonnet Coupling;Phys. Rev. Lett. 120, 131104 (2018) - Published 30 March 2018

8.   Jahed Abedi, Hannah Dykaar, and Niayesh Afshordi; Echoes from the abyss: Tentative evidence for Planck-scale structure at black hole horizons ;Phys. Rev. D 96, 082004 (2017) - Published 26 October 2017

9.   A. Hees, T. Do, A. M. Ghez, G. D. Martinez, S. Naoz, E. E. Becklin, A. Boehle, S. Chappell, D. Chu, A. Dehghanfar, K. Kosmo, J. R. Lu, K. Matthews, M. R.

111

Morris, S. Sakai, R. Schödel, and G. Witzel; Testing General Relativity with Stellar Orbits around the Supermassive Black Hole in Our Galactic Center; Phys. Rev. Lett. **118**, 211101 (2017) - Published 25 May 2017

10. Petr Hořava ; Spectral Dimension of the Universe in Quantum Gravity at a Lifshitz Point; Phys. Rev. Lett. 102, 161301 (2009); Published April 20, 2009

11. Kenji Hayashi and Takeshi Shirafuji ; Addendum to "New general relativity" ; Phys. Rev. D **24**, 3312 – Published 15 December 1981

12. Talmadge M. Davis and John R. Ray ; Ghost neutrinos in general relativity; Phys. Rev. D **9**, 331 (1974) - Published 15 January 1974

13. Patrick G. Whitman and Richard C. Burch ; Charged spheres in general relativity; Phys. Rev. D **24**, 2049 – Published 15 October 1981; Erratum Phys. Rev. D **25**, 1744 (1982)

14. Joseph Jacobson, Gunnar Björk, Isaac Chuang, and Yoshihisa Yamamoto ; Photonic de Broglie Waves ; Phys. Rev. Lett. **74**, 4835 (1995) - Published 12 June 1995

15. Bogeun Gwak and Bum-Hoon Lee ; Instability of rotating anti–de Sitter black holes ; Phys. Rev. D **91**, 064020 – Published 9 March 2015

16. Raphael Bousso and Stephen Hawking ; Erratum: Lorentzian condition in quantum gravity [Phys. Rev. D 59, 103501 (1999)] ; Phys. Rev. D **60**, 109903 (1999) - Published 8 October 1999

17   A. Steane, P. Szriftgiser, P. Desbiolles, and J. Dalibard ; Phase Modulation of Atomic de Broglie Waves ; Phys. Rev. Lett. **74**, 4972 – Published 19 June 1995

18   S. B. Cahn, A. Kumarakrishnan, U. Shim, T. Sleator, P. R. Berman, and B. Dubetsky ; Time-Domain de Broglie Wave Interferometry ; Phys. Rev. Lett. **79**, 784 – Published 4 August 1997

19   J. L. Synge ; Primitive Quantization in the Relativistic Two-Body Problem ; Phys. Rev. **89**, 467 – Published 15 January 1953

20   H. Jehle, Flux Quantization and particle Physics, Phys. Rev. D6 (1972) 441 – 457

21   Osung Kwon, Young-Sik Ra, and Yoon-Ho Kim ; Observing photonic de Broglie waves without the maximally-path-entangled $|N,0\rangle+|0,N\rangle$ state ; Phys. Rev. A **81**, 063801 (2010) - Published 1 June 2010

22   V. Krachmalnicoff, J.-C. Jaskula, M. Bonneau, V. Leung, G. B. Partridge, D. Boiron, C. I. Westbrook, P. Deuar, P. Ziń, M. Trippenbach, and K. V. Kheruntsyan ; Spontaneous Four-Wave Mixing of de Broglie Waves: Beyond Optics ; Phys. Rev. Lett. **104**, 150402 (2010) - Published 15 April 2010

23   Andrey Turlapov, Alexei Tonyushkin, and Tycho Sleator ; Talbot-Lau effect for atomic de Broglie waves manipulated with light ; Phys. Rev. A **71**, 043612 (2005) - Published 25 April 2005

24   Jakob Petersen, Eli Pollak, and Salvador Miret-Artes ; Alberto Nicolis and Riccardo Penco ; Mutual interactions of phonons, rotons, and gravity ; Phys. Rev. B **97**, 134516 (2018) - Published 18 April 2018

25   Quantum threshold reflection is not a consequence of a region of the long-range attractive potential with rapidly varying de Broglie wavelength ; Phys. Rev. A **97**, 042102 (2018) - Published 3 April 2018

113

26   Jing, H., Jiang, Y. & Deng, Y.; Quantum superchemistry of de Broglie waves: New wonderland at ultracold temperature ; Front. Phys. China (2011) 6: 15. https://doi.org/10.1007/s11467-010-0155-y

27   Donald H Kobe ; Quantum power in de Broglie–Bohm theory ; Journal of Physics A: Mathematical and Theoretical, Volume 40 - Number 19 , Published 24 April 2007

28   Rodewald, J., Haslinger, P., Dörre, N. et al.; New avenues for matter-wave-enhanced spectroscopy ; Appl. Phys. B (2017) 123: 3. https://doi.org/10.1007/s00340-016-6573-y

29.   J. W. Vegt, A Continuous Model of Matter based on AEONs, Physics Essays ,1995, Volume 8, Number 2, 201-224   A Continuous model of Matter (*DOI: 10.13140/RG.2.2.25149.77281*).

30.   J. W. Vegt, Annales Fondation Louis de Broglie, The Maxwell-Schrödinger-Dirac Correspondence in Auto Confined Electromagnetic Fields, Annales Fondation Louis de Broglie, 2002, Volume 27, Number 1,

31.   J. W. Vegt, A particle Free Model of Matter based on Electromagnetic Self-Cofinement, Annales Fondation Louis de Broglie, 1996, January.

32.   J. M. Maldacena , Black Holes in String Theory, Princeton University, arxiv.org/abs/hep-th/960723533.

33.   V. C. de Andrade and J. G. Pereira, Gravitational Lorentz force and the description of the gravitational interaction, Phys. Rev. D **56**, 468

34.   Mohr, P.J.; Taylor, B.N.; Newell, D.B. (2006). "CODATA recommended values of the fundamental physical constants". Reviews of Modern Physics. 80(2): 633–730. arXiv:0801.0028, Bibcode:2008RvMP...80..633M. doi:10.1103/RevModPhys.80.633.

35.   J. M. Maldacena , Black Holes in String Theory, Princeton University, arxiv.org/abs/hep-th/960723533.

36.   V. C. de Andrade and J. G. Pereira, Gravitational Lorentz force and the description of the gravitational interaction, Phys. Rev. D **56**, 468

37.   Mohr, P.J.; Taylor, B.N.; Newell, D.B. (2006). "CODATA recommended values of the fundamental physical constants". Reviews of Modern Physics. 80(2): 633–730. arXiv:0801.0028, Bibcode:2008RvMP...80..633M. doi:10.1103/RevModPhys.80.633.

38.   William A. Hiscock, Phys. Rev. D **31**, 3288 – Published 15 June 1985

39.   T. Degrand, L. Jaffe, K Johnson, J. Kiskis, Masses and other parameters of the light Hadrons, Physical Review D: Particles and Fields, 12(7), October 1975, 2060-2076

40   Volodymyr Krasnoholovets, Motion of a Relativistic Particle and the Vacuum, Physics Essays, vol 10, no 3, 1997, 407-416, arXiv:quant-ph/9903077

41   H. Jehle, Flux Quantization and fractional charges of quarks, Phys. Rev. D11(1975) 2147

42   W.G.V. Rosser, Classical Electromagnetism via Relativity (Butterworths, London, 1968), p. 134.

43.   Brando Bellazzini, Francesco Riva, Javi Serra, and Francesco Sgarlata; Beyond Positivity Bounds and the Fate of Massive Gravity ; Phys. Rev. Lett. **120**, 161101 – Published 17 April 2018

44.   Bao-Fei Li, Parampreet Singh, and Anzhong Wang; Towards cosmological dynamics from loop quantum gravity ;Phys. Rev. D **97**, 084029 – Published 17 April 2018

www.ingramcontent.com/pod-product-compliance
Lightning Source LLC
Chambersburg PA
CBHW040808200526
45159CB00022B/49